SpringerBriefs in Materials

More information about this series at http://www.springer.com/series/10111

Mohit Gupta

Design of Thermal Barrier Coatings

A Modelling Approach

 Springer

Mohit Gupta
University West
Trollhättan, Sweden

ISSN 2192-1091   ISSN 2192-1105   (electronic)
SpringerBriefs in Materials
ISBN 978-3-319-17253-8   ISBN 978-3-319-17254-5   (eBook)
DOI 10.1007/978-3-319-17254-5

Library of Congress Control Number: 2015937967

Springer Cham Heidelberg New York Dordrecht London

Printed on acid-free paper

Springer International Publishing AG Switzerland is part of Springer Science+Business Media
(www.springer.com)

*Dedicated to my beloved parents
Smt. (Late) Raj Kumari Gupta and Shri Ram
Niwas Gupta*

Preface

Thermal barrier coatings (TBCs) are now key elements in the design of advanced gas turbines. TBCs are used in a wide variety of modern gas turbine applications such as power generation, marine and aero engines. Applications of TBCs result in higher gas turbine efficiency as well as enhanced lifetime of components due to the thermal protection provided by TBCs. An extensive research is being carried out in this field to continue the implementation of new and advanced coating systems to achieve even lower emissions as well as fuel costs.

A high performance TBC would exhibit low thermal conductivity, high strain tolerance and long lifetime. Thus, these three properties need to be optimised by controlling the microstructure defects which control the thermal–mechanical properties of TBCs as well as the topcoat–bondcoat interface roughness which controls the lifetime of TBCs. The purpose of this book is to describe a design methodology which could be implemented to obtain an optimised TBC to be used for next generation gas turbines.

Chapter 1 introduces the topic in detail and describes the scope of this book. A general background knowledge about the processing technology and materials in TBCs is given in Chap. 2. The characteristics of TBCs in terms of microstructure, properties and failure mechanisms are discussed in Chap. 3. The experimental methods commonly used to characterise TBCs are described briefly in Chap. 4.

Chapter 5 describes a modelling approach to design the thermal–mechanical properties of TBCs which consists of real microstructure images as well as artificially created images, and the results are obtained by using that approach. Chapter 6 describes a modelling approach to design the topcoat–bondcoat interface roughness in TBCs which consists of real interface roughness profiles, and the results are obtained by implementing that approach. Chapter 7 describes a diffusion-based modelling approach to study oxide formation in TBCs during service conditions, and the results are obtained by implementing that approach. The conclusions from the modelling approach described in this book are given in the last chapter as a methodology to design an optimised TBC.

This work was performed at the Production Technology Centre (PTC), Trollhättan, as a part of the Thermal Spray research group at University West. The major portion of this work was done during the doctoral studies of the author. The author would like to express his gratitude to his main supervisor during his doctoral studies, Prof. Per Nylén, for his guidance, great support and valuable suggestions during this work.

Trollhättan, Sweden   Mohit Gupta
25th February 2015

Contents

About the Author

Mohit Gupta comes from Lucknow in India and received his bachelor's degree in mechanical engineering in 2009 from Indian Institute of Technology Kanpur, India. He received his master's degree in mechanical engineering in 2010 and doctoral degree in production technology in 2015 from University West, Sweden. He is currently employed as a university lecturer at University West. His research interests are finite element modelling, plasma spraying and solid oxide fuel cells.

Abbreviations

2D	Two-dimensional
3D	Three-dimensional
APS	Atmospheric plasma-sprayed
CAD	Computer-aided design
CFD	Computational fluid dynamics
CMAS	Calcium–magnesium–aluminosilicate
CSN	Chromia, spinel and nickel oxide
CTE	Coefficient of thermal expansion
DoE	Design of experiments
DSC	Differential scanning calorimetry
EB-PVD	Electron beam-physical vapour deposition
ECP	Ex situ coating property
FDM	Finite difference method
FEM	Finite element method
FOD	Foreign object damage
HVOF	High velocity oxy-fuel
LFA	Laser flash analysis
LOM	Light optical microscopy
LPPS	Low pressure plasma spraying
Micro-CT	Micro-computed tomography
MIP	Mercury intrusion porosimetry
OOF	Object-oriented finite element analysis
PS-PVD	Plasma spray-physical vapour deposition
SANS	Small-angle neutron scattering
SEM	Scanning electron microscope
SPPS	Solution precursor plasma spraying
SPS	Suspension plasma spraying
TBC	Thermal barrier coating
TCF	Thermal cyclic fatigue
TGO	Thermally grown oxide

VPS	Vacuum plasma spraying
XMT	X-ray microtomography
XRD	X-ray diffraction
YSZ	Yttria-stabilised zirconia

Chapter 1
Introduction

Thermal barrier coating systems (TBCs) are widely used in modern gas turbine engines in power generation, marine and aero engine applications to lower the metal surface temperature in combustor and turbine section hardware. The application of TBCs can provide increased engine performance/thrust by allowing higher gas temperatures or reduced cooling air flow, and/or increased lifetime of turbine blades by decreasing metal temperatures. TBC is a duplex material system consisting of an insulating ceramic topcoat layer and an inter metallic bondcoat layer. It is designed to serve the purpose of protecting gas turbine components from the severe thermal environment, thus improving the efficiency and at the same time decreasing unwanted emissions. Turbine entry gas temperatures can be higher than 1,500 K with TBCs providing a temperature drop of even higher than 200 K across them [1]. TBCs were first successfully tested in the turbine section of a research gas turbine engine in the mid-1970s [2].

Improvement in the performance of TBCs remains a key objective for further development of gas turbine applications. A key objective for such applications is to maximise the temperature drop across the topcoat, thus allowing higher turbine entry temperatures and thus higher engine efficiencies. This comes with the requirement that the thermal conductivity of the ceramic topcoat should be minimised and also that the value should remain low during prolonged exposure to service conditions. In addition to this, longer lifetime of TBCs compared to the state-of the art is of huge demand in the industry. In case of land-based gas turbines, a lifetime of around 40,000 h is desired. Therefore, substantial research efforts are made in these areas as TBCs have become an integral component of most gas turbines and are a major factor affecting engine efficiency and durability.

The coating microstructures in TBC applications are highly heterogeneous, consisting of defects such as pores and cracks of different sizes. The density, size and morphology of these defects determine the coating's final thermal and mechanical properties and the service lives of the coatings [3–6]. To achieve a low thermal conductivity and high strain-tolerant TBC with a sufficiently long lifetime, an optimisation

© The Author(s) 2015
M. Gupta, *Design of Thermal Barrier Coatings*, SpringerBriefs in Materials,
DOI 10.1007/978-3-319-17254-5_1

between the distribution of pores and cracks is required, thus making it essential to have a fundamental understanding of microstructure–property relationships in TBCs to produce a desired coating.

Failure in atmospheric plasma-sprayed (APS) TBCs during thermal cyclic loading is often within the topcoat near the interface, which is a result of thermo-mechanical stresses developing due to thermally grown oxide (TGO) layer growth and thermal expansion mismatch during thermal cycling. These stresses induce the propagation of pre-existing cracks in as-sprayed state near the interface finally leading to cracks long enough to cause spallation of the topcoat [7–14]. The interface roughness, although essential in plasma-sprayed TBCs for effective bonding between topcoat and bondcoat, creates locations of high stress concentration. Therefore, the understanding of fundamental relationships between interface roughness and induced stresses, as well as their influence on lifetime of TBCs, is of high relevance.

The traditional methodology to optimise the coating microstructure is by under-taking an experimental approach. In this approach, certain set spray parameters are chosen based on prior experience with the equipment and process, or as is the case quite often, spray parameters from a factorial design experiments are chosen. Thereafter, coatings are deposited using these parameters and evaluated by different testing methods. This procedure could be iterated until the specified performance parameters are obtained. As it might be apparent, this experimental approach is extremely time consuming and expensive, apart from the drawback that it does not enhance our knowledge of quantitative microstructure–property relationships.

Therefore, the derivation of microstructure–property relationships by simulation would be an advantage. Simulation approach, apart from being time saving and cost effective, is highly useful for the establishment of quantitative microstructure–property relationships. New coating designs can be developed and analysed with the help of simulation in a much easier manner compared to the experimental approach. Another advantage of using simulation is that the individual effects of microstructural features (such as defects and roughness profiles) could be artificially separated and analysed to study their effect on properties of TBCs independently which is not possible by experimental methods. However, it must be noted here that the relationships between process parameters and microstructure have to be established via an experimental approach [4, 6]. A simulation approach is schematically described in Fig. 1.1, showing the steps required to achieve a high performance TBC requiring low thermal conductivity and long lifetime. It should be noted that the microstructural parameters for modelling are obtained through the link with experimental spray parameters during the optimisation routine via process maps as described further in Sect. 2.4.

A two-step approach has been described in this book to design an optimised TBC by modelling:

1. *Design microstructure defects in topcoat*: As the thermal–mechanical properties of TBCs are highly dependent on coating microstructure, the microstructure defects can be designed to optimise the topcoat so as to achieve low thermal conductivity and high strain tolerance which would imply long lifetime.

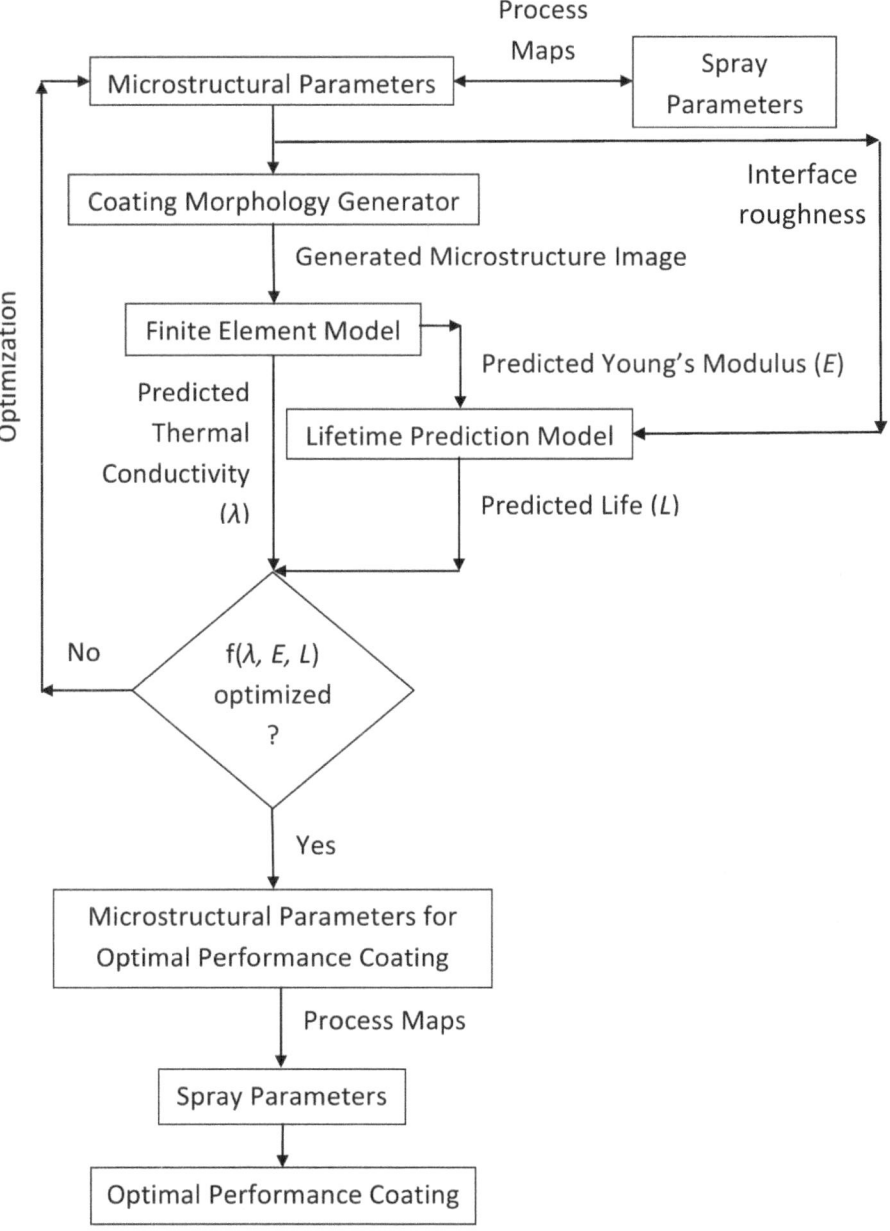

Fig. 1.1 Block diagram showing the steps required to obtain a high performance TBC

2. *Design topcoat–bondcoat interface*: As the crack propagation in TBCs near topcoat–bondcoat interface is highly dependent on the interface topography, the interface can be designed to optimise the topography so as to reduce thermomechanical stresses leading to crack propagation and thus enhance lifetime.

1.1   Scope and Limitations

The present study is general in itself as it is based on analysis of microstructure images and roughness profiles and is not dependent on material or equipment used to fabricate them. However, other factors might have to be considered if this study is applied to other materials or coating applications.

The limitations in this work can be broadly divided into the following categories:

1. *Process*
   The study was limited to APS for topcoats and mainly high velocity oxy-fuel (HVOF) spraying for bondcoats using powder feedstock. The effect of varying bondcoat spray parameters on bondcoat microstructure and surface roughness was not considered.
2. *Material*
   Only one topcoat material, namely zirconia, was considered with different stabilisers such as yttria and dysprosia. NiCoCrAlY was the only bondcoat material considered in this study.
3. *Experimental evaluation techniques*
   The experimental analysis performed in this study was limited to the scope of the used technique. Microstructure was evaluated with light optical microscopy (LOM) and scanning electron microscope (SEM) which could be incapable of detecting all fine pores and cracks present in the microstructure. Lifetime testing was limited to thermal cyclic fatigue (TCF) testing.
4. *Modelling*
   The effect of changing chemistry was not considered in this study. The microstructure considered in this study consisted of only the defect morphology. The thermal–mechanical properties modelled in this study were thermal conductivity and Young's modulus. Lifetime assessment considered in the modelling work was based on TCF testing.

Most of the material properties used in the modelling work were considered to be temperature independent even though the properties could change significantly over the wide range of temperatures considered. As the focus was set on qualitative analysis rather than on predicting exact values, this assumption was considered to be valid in this case. The effect of radiation was not considered when predicting thermal conductivity as it can be assumed to be scattered due to the porous microstructure.

Two-dimensional (2D) domain was considered in several models used to determine coating properties which could affect the final values, though it can be used effectively for comparative purposes. Virtually designed microstructures were limited by the scope of the software used. Focus was placed on the individual influence of microstructural features on the final properties of the coating.

The effect of coating thickness was not considered in the stress analysis model. The failure mechanism considered in the modelling work was limited to spallation of coating due to thermo-mechanical stresses; other failure mechanisms such as erosion were not considered.

References

1. Golosnoy IO, Cipitria A, Clyne TW (2009) Heat transfer through plasma-sprayed thermal barrier coatings in gas turbines: a review of recent work. J Therm Spray Technol 18(5–6): 809–821
2. Miller RA (1997) Thermal barrier coatings for aircraft engines: history and directions. J Therm Spray Technol 6(1):35–42
3. McPherson R (1981) The relationship between the mechanism of formation, microstructure and properties of plasma-sprayed coatings. Thin Solid Films 83(3):297–310
4. Friis M (2002) A methodology to control the microstructure of plasma sprayed coatings. Ph.D. thesis, Lund University, Sweden
5. Vassen R, Träeger F, Stöver D (2004) Correlation between spraying conditions and microcrack density and their influence on thermal cycling life of thermal barrier coatings. J Therm Spray Technol 13(3):396–404
6. Sampath S, Srinivasan V, Valarezo A, Vaidya A, Streibl T (2009) Sensing, control, and in situ measurement of coating properties: an integrated approach toward establishing process-property correlations. J Therm Spray Technol 18(2):243–255
7. Evans HE (2011) Oxidation failure of TBC systems: an assessment of mechanisms. Surf Coat Technol 206:1512–1521
8. Hille TS, Turteltaub S, Suiker ASJ (2011) Oxide growth and damage evolution in thermal barrier coatings. Eng Fract Mech 78:2139–2152
9. Kim D-J, Shin I-H, Koo J-M, Seok C-S, Lee T-W (2010) Failure mechanisms of coin-type plasma-sprayed thermal barrier coatings with thermal fatigue. Surf Coat Technol 205: S451–S458
10. Vaßen R, Giesen S, Stöver D (2009) Lifetime of plasma-sprayed thermal barrier coatings: comparison of numerical and experimental results. J Therm Spray Technol 18(5–6):835–845
11. Trunova O, Beck T, Herzog R, Steinbrech RW, Singheiser L (2008) Damage mechanisms and lifetime behavior of plasma sprayed thermal barrier coating systems for gas turbines—part I: experiments. Surf Coat Technol 202:5027–5032
12. Chen WR, Wu X, Marple BR, Patnaik PC (2006) The growth and influence of thermally grown oxide in a thermal barrier coating. Surf Coat Technol 201:1074–1079
13. Zhu D, Choi SR, Miller RA (2004) Development and thermal fatigue testing of ceramic thermal barrier coatings. Surf Coat Technol 188–189:146–152
14. Schlichting KW, Padture NP, Jordan EH, Gell M (2003) Failure modes in plasma-sprayed thermal barrier coatings. Mater Sci Eng A 342:120–130

Chapter 2
Background

2.1 Thermal Spraying

Thermal spraying is a branch of surface engineering processes in which metallic or non-metallic coating material (in powder, wire or rod form) is heated to a molten or semi-molten state and then propelled towards a prepared surface by either process gas or atomisation jets. These particles adhere to the surface and build up to form a coating. The workpiece on which the coating is deposited, or the substrate, remains unmelted.

The credit for inventing thermal spraying process belongs to M. U. Schoop (Zurich, Switzerland) [1] who, along with his associates, developed equipment and techniques for producing coatings using molten and powder metals in the early 1900s. The process was initially called "Metallization." In 1908, Schoop patented the electric arc spray process. In 1939, Reinecke introduced the first plasma spraying process. Advancements in thermal spray equipment technology saw much higher pace after the 1950s. Different variations of thermal spray technique exist today, such as flame spraying, HVOF, APS and vacuum plasma spraying (VPS).

A major advantage of thermal spraying is that it can be used to deposit a wide variety of materials without a significant heat input. In theory, any material that melts without decomposing can be used for spraying without any undue distortion of the part. A major disadvantage is that thermal spraying is a "line of sight" process, although new processes such as plasma spray-physical vapour deposition (PS-PVD) allow even shadowed areas to be coated [2].

2.1.1 Atmospheric Plasma Spraying

Plasma is an electrically conductive gas containing charged particles. When a gas is excited to high energy levels, atoms lose hold of some of their electrons and become ionised producing plasma containing electrically charged particles (ions and electrons)

© The Author(s) 2015
M. Gupta, *Design of Thermal Barrier Coatings*, SpringerBriefs in Materials,
DOI 10.1007/978-3-319-17254-5_2

with temperatures ranging up to 20,000 K. The plasma generated for plasma spray process usually incorporates one or a mixture of argon, helium, nitrogen and hydrogen. The advantage of plasma flame is that it supplies large amounts of energy through dissociation of molecular gases to atomic gases and ionisation.

A typical plasma spray process can be described in the following steps:

- First, a gas flow mixture (H_2, N_2, Ar) is introduced between a water-cooled copper anode and a tungsten cathode.
- A high intensity DC electric arc passes between cathode and anode and is ionised to form a plasma to reach extreme temperatures.
- The coating material in the form a fine powder conveyed by carrier gas (usually argon) is introduced into the plasma plume formed due to the flow gases via an external powder port and is heated to the molten state.
- The compressed gas propels the molten particles towards the substrate with particle velocities ranging from 200 to 800 m/s.

Figure 2.1 shows a photograph taken during the plasma spray process indicating the different components of the spraying unit.

Materials suitable for plasma spraying include zinc, aluminium, copper alloys, tin, molybdenum, some steels, and numerous ceramic materials [1]. The advantage of using plasma spray process compared to combustion processes is that it can spray materials with very high melting points (refractory metals like tungsten and ceramics like zirconia). On the other hand, it requires higher cost and increases the process complexity.

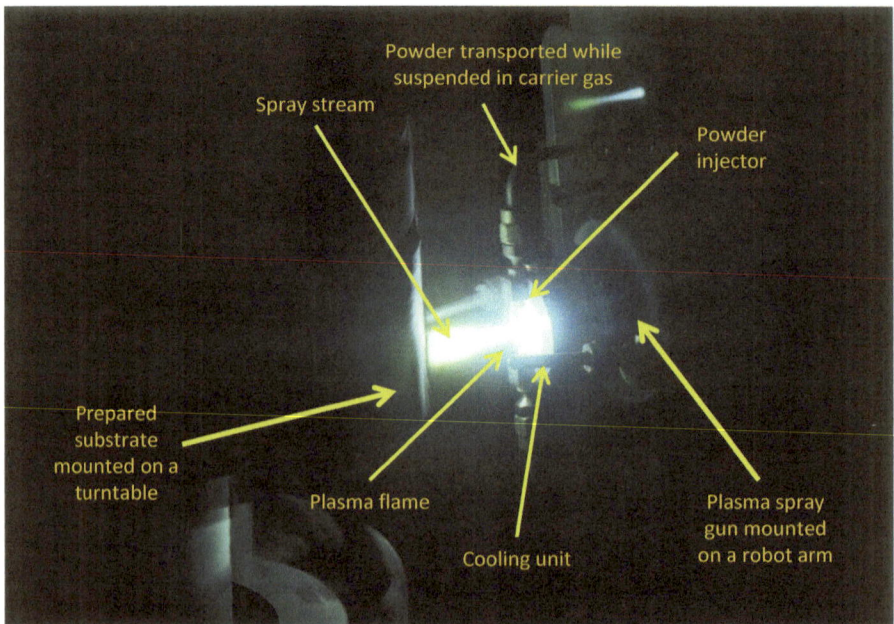

Fig. 2.1 A photograph taken during atmospheric plasma spraying process

In some cases, it could be advantageous to perform plasma spray process under controlled environment under low pressure or vacuum, correspondingly termed as low pressure plasma spraying (LPPS) or VPS. The use of controlled environment could improve the coating quality due to reduced oxidation while spraying. However, these processes increase the costs for the set-up equipment as well as processing times, thus leading to significantly increased overall costs.

2.1.2 High Velocity Oxy-Fuel Spraying

In high velocity oxy-fuel (HVOF) spraying, a mixture of a fuel gas (such as hydrogen, propane and propylene) and oxygen is ignited in a combustion chamber at high pressures and the combustion gases are accelerated through a long de Laval (convergent–divergent) nozzle to generate a supersonic jet with very high particle speeds. This spray process generates extremely dense and well-bonded coatings.

The coatings usually sprayed by HVOF process are hard cermets like WC/Co or Cr_2C_3/NiCr or MCrAlY (M = Ni and/or Co) applied as bondcoats to aircraft turbine blades [1]. The advantage of using HVOF, apart from being suitable for making dense coatings, is that the coatings contain few oxides due to the low process temperature making it very attractive for TBC bondcoat applications.

2.1.3 Liquid Feedstock Plasma Spraying

The limitation of minimum particle size which could be used for APS led to the development of plasma spray technology based on using liquid feedstock, mainly in the form of suspension or solution, correspondingly known as suspension plasma spraying (SPS) or solution precursor plasma spraying (SPPS). Since powders below 5 μm in size are difficult to feed and inject into the plasma torch, nanostructured coatings are obtained by dispersing/dissolving nano- or sub-micrometric powder in a liquid media to create a suspension/solution, respectively, and using it as a feedstock [3]. Suspensions are based on either water or an organic solvent in the form of alcohol which is typically ethanol. Solutions are typically made of nitrates or chlorides which are oxidised during the spray process forming an oxide particle that forms the coating [3].

2.2 Thermal Barrier Coatings

A typical TBC (schematically shown in Fig. 2.2) consists of an intermediate metallic bondcoat and a ceramic topcoat that provides the temperature drop across the coating. In addition to a low thermal conductivity, topcoats should also have phase stability during long-term high-temperature exposure and thermal cycling. The present

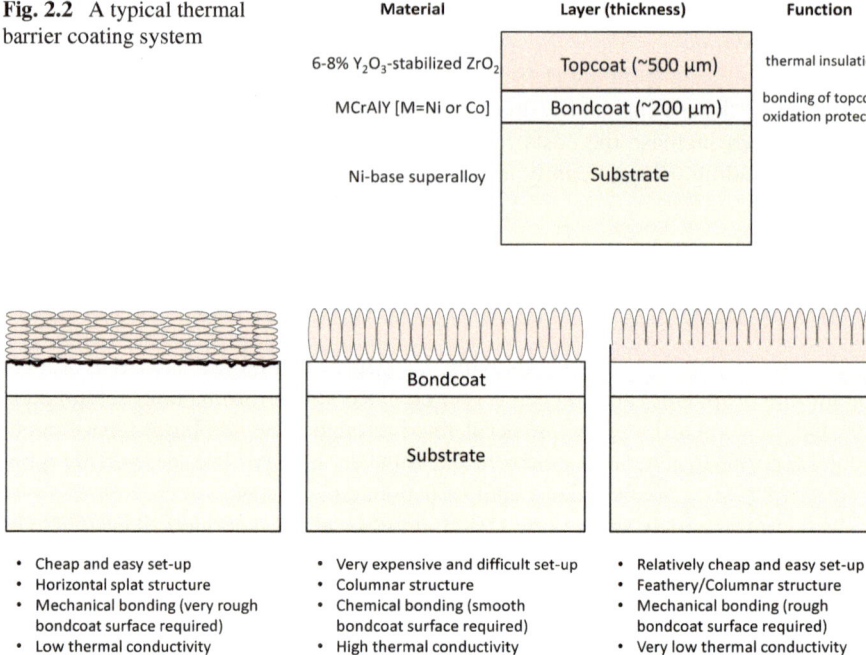

Fig. 2.2 A typical thermal barrier coating system

Material	Layer (thickness)	Function
6-8% Y_2O_3-stabilized ZrO_2	Topcoat (~500 μm)	thermal insulation
MCrAlY [M=Ni or Co]	Bondcoat (~200 μm)	bonding of topcoat, oxidation protection
Ni-base superalloy	Substrate	

- Cheap and easy set-up
- Horizontal splat structure
- Mechanical bonding (very rough bondcoat surface required)
- Low thermal conductivity
- Short lifetime

- Very expensive and difficult set-up
- Columnar structure
- Chemical bonding (smooth bondcoat surface required)
- High thermal conductivity
- Very long lifetime

- Relatively cheap and easy set-up
- Feathery/Columnar structure
- Mechanical bonding (rough bondcoat surface required)
- Very low thermal conductivity
- Relatively long lifetime

Fig. 2.3 A simplified comparison of APS (*left*), EB-PVD (*centre*) and SPS (*right*) TBC properties

state-of-the-art topcoat material is a 6–8 wt% yttria-stabilised zirconia (YSZ) ceramic applied usually with plasma spray on a bondcoat of NiCoCrAlY [1]. YSZ has a low thermal conductivity compared to other ceramics such as alumina. Also, it is durable and chemically stable with a high melting point which makes it a good choice. Further benefits of this material are discussed in detail in Sect. 2.5.1.

The ceramic top layer is typically applied by APS, electron beam-physical vapour deposition (EB-PVD) or SPS. Figure 2.3 summarises a simplified comparison of APS, EB-PVD and SPS TBCs. The main purpose of the topcoat is to provide thermal protection. In addition, it should be susceptible to the thermo-mechanical stresses arising during the operating conditions.

The bondcoat provides improved bonding strength between the substrate and the topcoat. It also reduces the interface stresses arising due to the difference in coefficients of thermal expansion (CTEs) of ceramic topcoats and metallic substrates. As at high temperatures the porous ceramic topcoat is transparent to the flow of oxygen and the exhaust gases, an aluminium-enriched bondcoat composition is used to provide a slow growing, adherent aluminium oxide film known as TGO. TGO layer provides oxidation protection to the substrate as alumina has very low diffusion coefficients for both oxygen and metal ions.

2.3 Coating Formation

Plasma-sprayed coatings are built up particle by particle. Each individual molten or semi-molten particle impacts the substrate surface and flattens, adheres and solidifies to form a lamellae structure called splat. Unmelted particles are bounced back from the substrate reducing the deposition efficiency of the coating. When a spherical liquid droplet strikes a flat surface with a high impact velocity, it tries to flatten to a disc but the radially flowing thin sheet of liquid becomes unstable and disintegrates at the edges to form small droplets. This process is interrupted by rapid solidification in the case of plasma spray as the substrate is well below the melting point of the droplet, with a cooling rate as high as 10^6 K s^{-1} [4]. Major part of the heat energy from the particle is transferred to the substrate by conduction, and solidification starts at the interface between the particle and the substrate. Heat transfer due to convection and radiation makes a small contribution at the conditions of plasma spray process [5].

Small voids exist between the splats, which can be formed due to incomplete bonding between splats owing to lack of adhesion, relaxation of the residual stresses mainly during the rapid cooling of the splat or trapped gases. The real area of contact between splats is around 20 % [4], due to which the properties of the coatings, such as mechanical, thermal and electrical properties, are very different compared to the sprayed bulk material. The major factors influencing the structure of a coating are the temperature, velocity and size distribution of the incident particles, apart from substrate temperature and roughness [6].

2.4 Process Parameters

Plasma spray process is influenced by a number of process parameters. The dependence of microstructure on so many parameters has also been an enigma in terms of process control. A vast number of parameters need to be monitored and controlled, and several others are very difficult to control, such as electrode wear, humidity in surrounding air and fluctuations in controllable parameters. Several studies have been performed in the past emphasising on examination of process–structure–property relationships based on process maps to overcome this drawback to some extent [5, 7–9]. These process maps, however, are restricted to the specific spray gun, powder and all other parameters, except the ones varied to conduct the study, which makes implementation of the process maps from one gun to another rather difficult. Some of the process parameters in plasma spraying are presented in Fig. 2.4.

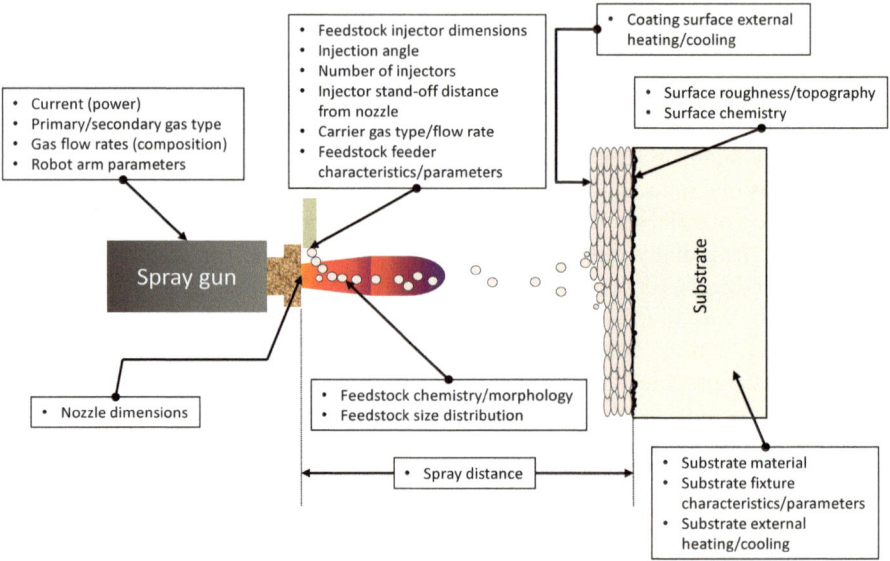

Fig. 2.4 Plasma spray process parameters

2.5   Coating Materials for TBCs

2.5.1   Topcoat

The ceramic topcoat provides thermal insulation to the substrate underneath. Some of the basic properties to be exhibited by the material used as topcoat are (1) low thermal conductivity, (2) high melting point and (3) phase stability. The material most widely used as topcoat is 6–8 wt% YSZ due to its low thermal conductivity, relatively high CTE and adequate toughness [10]. Apart from YSZ, other ceramics which are used as TBC materials are mullite, Al_2O_3, TiO_2, CeO_2 + YSZ, $La_2Zr_2O_7$, pyrochlores and perovskites [11].

Zirconia ceramics are one of the very few non-metallic materials which have good mechanical properties as well as electrical properties, apart from having low thermal conductivity. These properties are exhibited due to the particular crystal structure of ZrO_2, which is principally a fluorite-type lattice. Pure ZrO_2 has a monoclinic crystal structure at room temperature and undergoes phase transformations to tetragonal (at 1,197 °C) and cubic (at 2,300 °C) at increasing temperatures with a melting point of about 2,700 °C [11, 12]. Figure 2.5 shows the phase diagram for the zirconia–yttria system. The Zr^{4+} ions in the cubic ZrO_2 have very low coefficients of diffusion as they are very immobile which gives ZrO_2 a very high melting point and good resistance to both acids and alkalis. On the other hand, the O^{2-} ions are mobile

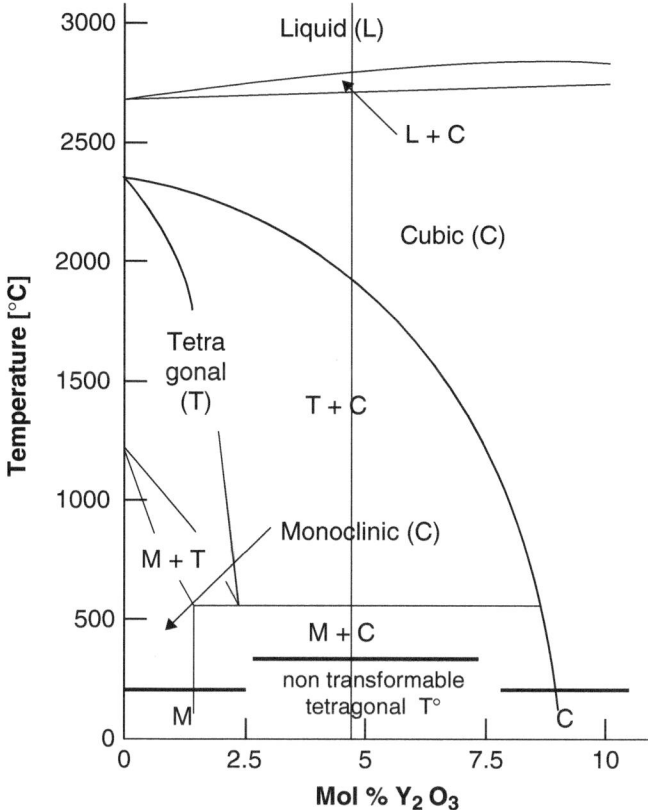

Fig. 2.5 Phase diagram for the zirconia–yttria system [13]

due to the presence of vacancies, which makes the cubic phase ion conducting and thus providing it good electrical properties. The main sources of ZrO_2 in nature are zircon ($ZrSiO_4$) and baddeleyite. Apart from TBCs, ZrO_2 has widespread applications in fuel cells, jewellery (cubic ZrO_2), oxygen sensors, electronics, etc.

The volume expansion which occurs due to the phase transformation from cubic (c) to tetragonal (t) to monoclinic (m) phases causes large stresses which will produce cracks in pure ZrO_2 upon cooling from high temperature. Thus, several oxides are added to stabilise the tetragonal and/or cubic ZrO_2 phases like yttria (Y_2O_3), ceria (Ce_2O_3), magnesium oxide (MgO) and calcium oxide (CaO). Specific addition of cations like Y^{3+}, Ca^{2+} into the ZrO_2 lattice causes them to occupy the Zr^{4+} positions in the lattice which results in the formation of anion vacancies to maintain charge equilibrium.

YSZ is preferred to CaO- or MgO-stabilised zirconia for TBCs as YSZ coatings have been proved to be more resistant against corrosion [11]. Also, YSZ coatings exhibit highest degree of resistance to coating failure due to spallation and an excellent thermal stability [14]. Its major disadvantage is the limited operating temperature (<1,473 K) for long-term application [11]. At higher temperatures, the metastable t'-tetragonal phase, which is the main phase present at the room temperature, transforms first to tetragonal and cubic $(t+c)$ phase and then to monoclinic (m) phase during cooling, which results in volume contraction and, consequently, the formation of cracks in the coating [11]. Also, as the YSZ coatings possess high concentration of vacancies, they facilitate oxygen transport at high temperatures which results in the oxidation of the bondcoat. This leads to the failure of TBCs due to spallation of the ceramic. The latter problem is solved by inducing the formation of oxidation-resistant TGO layer between the topcoat and bondcoat by using alloys based on NiAl with various additions such as Cr, Co, Pt, Y and Hf as bondcoat materials [10].

2.5.2   Bondcoat

The bondcoat provides oxidation protection of the substrate and improved adhesion of the topcoat. Some of the basic requirements from a bondcoat material are (1) resistance to interdiffusion with the substrate and (2) high creep strength with suitable ductility.

The typically used thermal sprayed bondcoat materials consist of a variety of MCrAlX alloys, where M = Ni and/or Co and X = Y, Hf and/or Si [15]. Nickel is added to enhance oxidation resistance while cobalt for corrosion resistance [16]. Aluminium is added to bondcoat to act as a local aluminium reservoir to provide a slow growing TGO α-alumina during service conditions to provide oxidation protection. Alumina is preferred to other oxides due to its low thermal diffusivity and better adherence [17]. Chromium is added to enhance oxidation and corrosion resistance [18]. Yttrium is added to provide protection from sulphur diffusion into the coating by acting as a gettering site and promoting formation of alumina as well as its adhesion [17].

2.5.3   Thermally Grown Oxides

The oxidation of bondcoat in service conditions results in the formation of a TGO layer near the topcoat–bondcoat interface. The major oxide formed due to the oxidation of a typically used MCrAlY bondcoat is alumina. Other oxides which are usually formed are chromia $((Cr,Al)_2O_3)$, spinel $(Ni(Cr,Al)_2O_4$, nickel oxide (NiO) and silica [19]. Chromia, spinel and nickel oxide are usually abbreviated as CSN in literature. The oxidation of bondcoat has been recognised as one of the major causes for TBC failure [19, 20]; it will be discussed in detail in Sect. 3.5.

References

1. Davis JR (ed) (2009) Handbook of thermal spray technology. ASM International, Materials Park
2. Niessen K, Gindrat M (2011) Plasma spray-PVD: a new thermal spray process to deposit Out of the vapor phase. J Therm Spray Technol 20(4):736–743
3. Fauchais P, Etchart-Salas R, Rat V, Coudert JF, Caron N, Wittmann-Teneze K (2008) Parameters controlling liquid plasma spraying solutions, sols, or suspensions. J Therm Spray Technol 17(1):31–59
4. McPherson R (1989) A review of microstructure and properties of plasma sprayed ceramic coatings. Surf Coat Technol 39–40:173–181
5. Friis M (2002) A methodology to control the microstructure of plasma sprayed coatings, Ph.D. thesis, Lund University, Sweden
6. Bianchi L, Leger AC, Vardelle M, Vardelle A, Fauchais P (1997) Splat formation and cooling of plasma-sprayed zirconia. Thin Solid Films 305(1–2):35–47
7. Vassen R, Träeger F, Stöver D (2004) Correlation between spraying conditions and microcrack density and their influence on thermal cycling life of thermal barrier coatings. J Therm Spray Technol 13(3):396–404
8. Sampath S, Jiang XY, Matejicek J, Leger AC, Vardelle A (1999) Substrate temperature effects on splat formation, microstructure development and properties of plasma sprayed coatings part I: case study for partially stabilized zirconia. Mater Sci Eng A 272:181–188
9. Vaidya A, Srinivasan V, Streibl T, Friis M, Chi W, Sampath S (2008) Process maps for plasma spraying of yttria-stabilized zirconia: an integrated approach to design, optimization and reliability. Mater Sci Eng A 497:239–253
10. Evans AG, Clarke DR, Levi CG (2008) The influence of oxides on the performance of advanced gas turbines. J Eur Ceram Soc 28(7):1405–1419
11. Cao XQ, Vassen R, Stoever D (2004) Ceramic materials for thermal barrier coatings. J Eur Ceram Soc 24(1):1–10
12. Schafer U, Schubert H, Carle VM, Taffner U, Predel F, Petzow G (1991) Ceramography of high performance ceramics—description of materials, preparation, etching techniques and description of microstructures—part III: zirconium di oxide (ZrO_2). Pract Metallogr 28:468–483
13. Scott HG (1975) Phase relationships in the zirconia-yttria system. J Mater Sci 10(9):1527–1535
14. Alpérine S, Derrien M, Jaslier Y, Mévrel R. Thermal barrier coatings: the Thermal Conductivity challenge, Proceedings of the 85th Meeting of the AGARD Structures and Materials Panel, Oct 15–16, 1997 (Aalborg, Denmark), NATO-AGARD-R-823, 1998, p 1-1–1-10
15. Haynes JA, Ferber MK, Porter WD (2000) Thermal cycling behavior of plasma-sprayed thermal barrier coatings with various MCrAIX bond coats. J Therm Spray Technol 9(1):38–48
16. Curry N (2014) Design of thermal barrier coatings. Ph.D. thesis, University West, Sweden, No. 3
17. Evans AG, Mumm DR, Hutchinson JW, Meier GH, Pettit FS (2001) Mechanisms controlling the durability of thermal barrier coatings. Prog Mater Sci 46:505–553
18. Yuan K (2014) Oxidation and corrosion of new MCrAlX coatings: modelling and experiments, Ph.D. thesis, Linköping University, Sweden
19. Chen WR, Wu X, Marple BR, Nagy DR, Patnaik PC (2008) TGO growth behaviour in TBCs with APS and HVOF bond coats. Surf Coat Technol 202:2677–2683
20. Vaßen R, Giesen S, Stöver D (2009) Lifetime of plasma-sprayed thermal barrier coatings: comparison of numerical and experimental results. J Therm Spray Technol 18(5–6):835–845

Chapter 3
Characteristics of TBCs

The characteristics of TBCs are very different from that of the bulk material. Apart from the presence of several defects present in plasma-sprayed coatings, the splat interfaces present in coatings are rough, resulting in multiple contact points ranging from micrometre to nanometre scale. These contact points can have very different bonding characteristics. These features have a significant implication on the macroscale properties of TBCs.

3.1 Microstructure

As discussed earlier, the microstructure of APS ceramic coatings is significantly influenced by the process parameters. This influence results in a complex microstructure with various forms of porous features. Common features present in a ceramic coating are presented in Fig. 3.1. This image of coating cross section has been taken by a SEM using backscattered electron detector mode which is a common technique used for analysing the microstructures in TBCs.

As it can be observed in Fig. 3.1, the microstructure consists mainly of large porous features, commonly referred to as globular pores, or simply pores, and several long and narrow porous features, commonly referred to as delaminations and cracks. They are also called interlamellar (delaminations, horizontally oriented) and intralamellar (vertical cracks or those non-horizontally oriented). It could be difficult to differentiate between delaminations and cracks, especially at lower magnifications, and so they are quite often simply referred to as cracks. Some fine scale porosity and partially melted particles can also be noticed in the microstructure image.

Globular pores are formed due to incomplete packing of the deposited particles or due to defects in the structure. This effect is magnified in the case of the low energy spray parameters since in that case the particles are unmelted/partially melted and so they do not flatten out adequately on impact. Delaminations are formed due to incomplete bonding between consecutively deposited particles, usually during

© The Author(s) 2015
M. Gupta, *Design of Thermal Barrier Coatings*, SpringerBriefs in Materials,
DOI 10.1007/978-3-319-17254-5_3

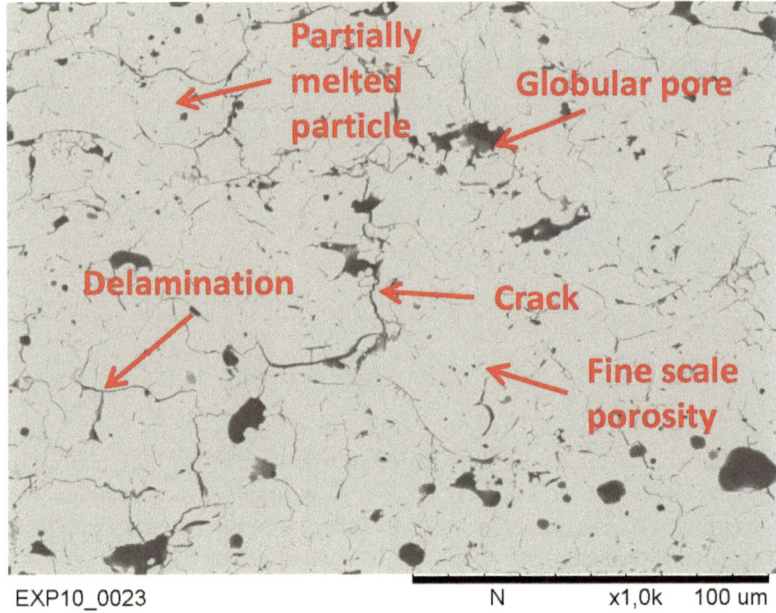

EXP10_0023   N   x1,0k   100 um

Fig. 3.1 SEM micrograph illustrating common features present in APS coating microstructures

successive passes of the spray gun. Cracks are formed due to the stresses developed within the coating when the sprayed particles cool down. The rapid cooling of particles results in large shrinkage which induces tensile stresses as the underlying material tries to prevent the shrinkage. These tensile stresses are released by the formation of cracks in the coating. Under appropriate spray conditions, vertical cracks in the coating can propagate to form long cracks orthogonal to the surface known as segmentation cracks. One such microstructure image is shown in Fig. 3.2.

The individual splats have a typical columnar grain structure caused by the directional solidification of the particles. This structure can be typically observed in a fracture surface of the coating as shown in Fig. 3.3. It can be observed that the columnar grains grow as usual along the direction of heat flow from top to bottom [2]. The grain structure and size depends on the processing conditions of the particles during spraying and can vary significantly during service conditions which could affect the coating properties [3].

Thermal–mechanical properties and lifetime of TBCs depend mainly on the various microstructural features present in the topcoat. A thermal sprayed YSZ coating has a thermal conductivity around $0.5–1$ W m^{-1} K^{-1}, as compared to the thermal conductivity of 2.5 W m^{-1} K^{-1} for bulk material [4]. A large amount of pores and cracks perpendicular to the heat flow will provide better insulation properties to the coating [5]. On the other hand, these horizontal cracks might propagate due to the thermal and mechanical stresses during operating conditions and eventually lead to failure of the coating by spallation [4]. Segmentation cracks increase the flexibility

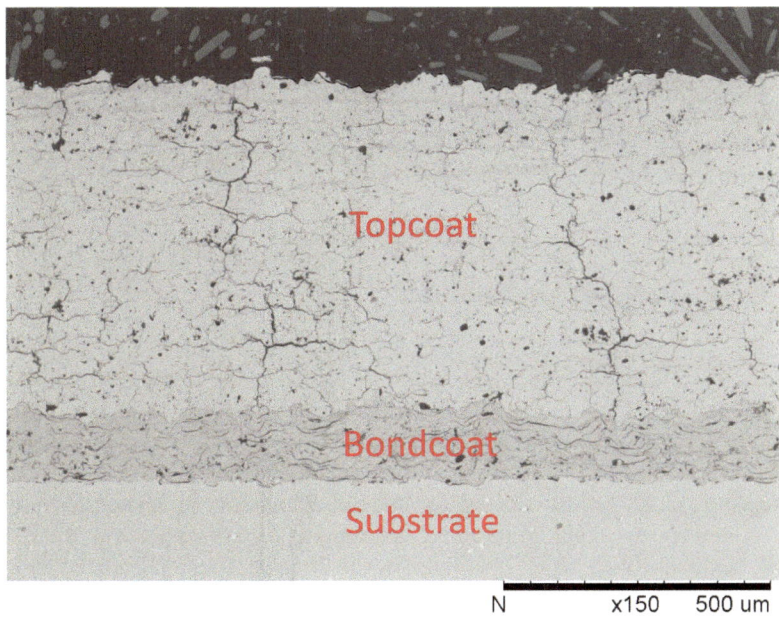

N   x150   500 um

Fig. 3.2 Microstructure image showing vertical cracks in the topcoat

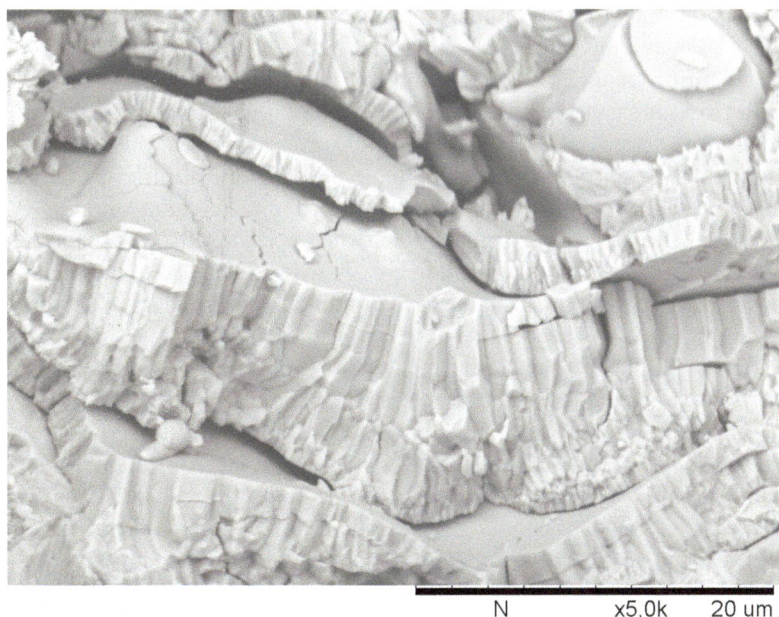

N   x5,0k   20 um

Fig. 3.3 A microstructure image showing the fracture surface of the coating in as-sprayed condition [1]

of the coating as they help in relaxing the residual stresses within the coating [6]. Again, on the other hand, these vertical cracks enhance the thermal conductivity of the coating as they allow the flow of high temperature gases thus increasing the heat transfer to the substrate [6]. To achieve a low thermal conductivity and high strain-tolerant TBC with a sufficiently long lifetime, an optimisation of microstructural features is required.

3.2 Heat Transfer Mechanism

3.2.1 General Theory

The theory of heat transfer in crystalline solids is described very well in the literature [7]. It has been briefly reviewed in other sources as well [6, 8–10].

Heat energy can be transferred by three mechanisms in crystalline solids—electrons, lattice waves (phonons) and electromagnetic waves (photons) [8]. The total thermal conductivity K, which is the sum of the three components, can be expressed in general form as [8]

$$\kappa = \frac{1}{3} \sum_{j=1}^{N} C_{p_j} v_j l_j \tag{3.1}$$

where Cp is the specific heat at constant pressure, N is the total number of energy carriers, v is the velocity of a given carrier (group velocity if the carrier is a wave) and l is the corresponding mean free path.

The electrons are capable of transferring energy only when they are free of interactions with the crystal lattice. This type of electrons is present only in metals and partly in metal alloys, especially at high temperatures. The electronic thermal conductivity part K_e is proportional to the product of the temperature and the electron mean free path, with v_e being independent of temperature [8]. The electron mean free path has two parts: residual mean free path, which is related to the scattering of electrons by defects, and intrinsic mean free path, which is related to the scattering of electrons by lattice vibrations. The residual mean free path is independent of temperature while intrinsic mean free path is directly proportional to temperature. At low temperatures, the electrons are mainly scattered by the defects, while as the temperature increases, the scattering mechanism by the phonons becomes more and more dominant. Therefore, the electronic component of thermal conductivity is proportional to temperature at low temperatures, becoming less dependent as temperature increases, and finally becoming independent of it at high temperatures [8].

Lattice thermal conduction, or heat energy transport by phonons, occurs in all types of solids, with the phenomenon being dominant in alloys at low temperatures and in ceramics. At low temperatures (T), the component of thermal conductivity related to heat transport by phonons, K_{ph}, may be represented by an exponential term

$\exp(T^*/T)$, whereas it is inversely proportional to temperature at high temperatures [8]. Here T^* is the characteristic temperature of the material which is generally proportional to the Debye temperature. K_{ph} may also be expressed as [10]

$$\kappa_{\mathrm{ph}} = \frac{1}{3}\int C_v \rho v l_p \tag{3.2}$$

where C_v is the specific heat at constant volume, ρ is the density, and l_p is the mean free path for scattering of phonons.

Heat energy transport by photon conduction (radiation) occurs especially at high temperatures in materials transparent to infrared radiation such as ceramics (over 1,200–1,500 K) and glasses (over 900 K) [8]. The radiative component of the thermal conductivity, K_r, can be expressed as [10]

$$\kappa_r = \frac{16}{3}\sigma_s n^2 l_r T^3 \tag{3.3}$$

where n is the refractive index, l_r is the mean free path for photon scattering (defined as the path length over which the intensity of radiation will reduce by a factor of $1/e$) and σ_s is the Stefan–Boltzmann constant.

In real crystal structures, scattering of phonons occurs due to their interaction with lattice imperfections in the ideal lattice, like vacancies, dislocations, grain boundaries, atoms of different masses and other phonons. Phonon scattering may also occur due to ions and atoms of different ionic radius as they distort the bond length locally, and thus elastic strain fields might be present in the lattice. The phonon mean free path l_p is defined as [10]

$$\frac{1}{l_p} = \frac{1}{l_i} + \frac{1}{l_{\mathrm{vac}}} + \frac{1}{l_{\mathrm{gb}}} + \frac{1}{l_{\mathrm{strain}}} \tag{3.4}$$

where l_i, l_{vac}, l_{gb} and l_{strain} are mean free paths associated with interstitials, vacancies, grain boundaries and lattice strain, respectively. The intrinsic lattice structure and the strain fields mainly affect the phonon mean free path in conventional materials, with the grain boundary term having the least effect.

Thermal conduction in gases depends on the molecular mean free path, λ, of the gases. λ is a function of temperature and pressure P and is proportional to T/P for ideal gas behaviour [6]. The thermal conductivity of a gas, K_g, in a constrained channel of length d_v can be expressed as [6, 11]

$$\kappa_g = \frac{\kappa_g^0}{1 + BT/(d_v P)} \tag{3.5}$$

where κ_g^0 is the normal (unconstrained) conductivity of the gas at the temperature concerned and B is a constant which depends on the gas type and the properties of the interacting solid surface [6].

3.2.2   Application to TBCs

In a real engine environment, TBCs protecting the substrate receive radiation which can be classified into the following two categories—far-field and near-field radiations [9]. Figure 3.4 shows the temperature distribution across a typical TBC system during service conditions from the hot gases in the combustor to the substrate. The operating temperature increase obtained due to the application of TBC is indicated in the figure.

Far-field radiation comes from the combustion gases which are at temperatures around 2,000 °C, having a spectral distribution same as that of a black body at that temperature. This far-field radiation makes a small contribution as the hot combustion gases have finite thickness and limited opacity and thus have reduced emissivity, which results in the far-field radiation being reduced by a geometrical factor.

Near-field radiation comes due to the layer of cooler gas, at around 1,200 °C, which is adjacent to the topcoat. The topcoat ceramic surface, which is at a similar temperature as this gas layer, emits radiation which can pass through the partially transparent ceramic topcoat to the metallic bondcoat and the substrate. This near-field radiation contributes to the thermal conductivity of the ceramic and can affect it significantly at elevated temperatures [9].

The three important factors influencing the heat conduction in typical thermal sprayed coatings are the dimensions of the grains, the impurities and the porosity [8].

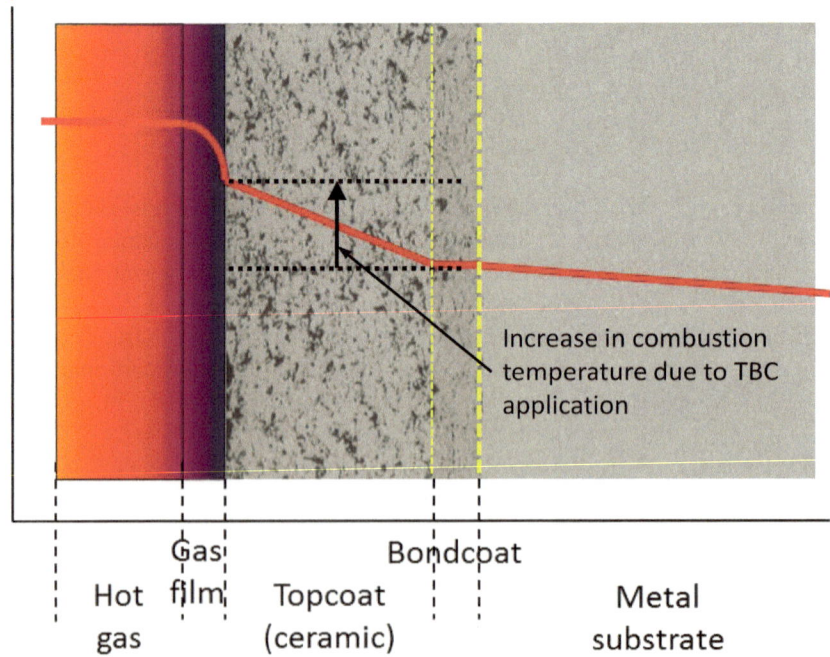

Fig. 3.4 Temperature distribution during typical service conditions across a typical thermal barrier coating system

The grain dimensions depend on the solidification conditions of sprayed liquid droplets, which depends mainly on spraying technique, cooling of the substrate and the thickness of the sprayed coating (particles solidifying on previously deposited layers have lower solidification rates and thus larger grain sizes) [8]. In case of oxides, however, the grain size might influence the phonon mean free path only at low temperatures, with the crystal structure being the major factor influencing mean free path rather than the grain size [8].

The typical impurities present in plasma-sprayed coatings are, besides oxides, the copper and tungsten particles coming from the electrodes of the torch and the sand blasted particles at the interface between coating and substrate. The effect of these impurities on thermal properties of coatings can be ignored if the electrodes are cleaned systematically and the substrate is cleaned after sand blasting [8].

The thermal conductivity of gas inside a pore is close to that of free gas if the dimensions of the pore are much larger than the mean free path ($L < 10\lambda$), but it can fall significantly below the free gas value for even moderately fine structures ($L < {\sim}1$ µm) [8]. Both pore thickness and gas pressure can affect the pore conductivity significantly, as well as the temperature [11]. Convective heat transfer within the pores can be neglected for plasma-sprayed TBCs, as it is only likely to be significant if the pores are large ($L > {\sim}10$ mm) [6]. Also, convective heat transfer through the porosity network can also be neglected, even though most porosity in TBCs is normally interconnected [6].

As zirconia-based ceramics are electronic insulators with electrical conductivity occurring only at high temperatures by oxygen ion diffusion, there is no contribution to thermal conductivity due to electrons. Thus heat transfer in zirconia takes place only by lattice vibrations (phonons) or radiation (photons) [10]. Adding yttria to zirconia modifies the lattice structure locally by introducing ion vacancies and generating local strain fields due to the incorporation of large dopant atoms, which results in lower intrinsic mean free path due to enhanced scattering of phonons and thus reduced thermal conductivity [10]. The radiative heat transfer part in zirconia-based plasma-sprayed TBCs becomes significant only at temperatures above ${\sim}1{,}500$ K and thus was neglected in the present modelling work [6]. The contribution from radiation depends on the radiation scattering length, which increases as the grain and pore structure coarsens and intersplat contact area increases as well as the coating thickness increases [6, 12].

3.3 Mechanical Behaviour

3.3.1 Stress Formation

The residual stresses in APS coatings in the as-sprayed condition arise due to two main factors [13, 14]:

1. **Quenching**: These stresses arise due to the rapid solidification of single particles during spraying while their contraction is restricted by the adherence to the substrate. Due to the temperature difference between the substrate and the particles, tensile stresses are generated in the particles known as quenching stresses.

2. **Thermal mismatch**: These stresses arise due to mismatch in CTE of the coating and the substrate during cooling after spraying. As ceramic topcoats have much lower CTE than the metallic substrates, the CTE mismatch results in compressive stresses in the coating as the substrate shrinks more during cooling.

The stress state in the coating is due to the combination of the above two factors leading to stress generation in APS coatings. In addition, several other factors could also attribute to the final stress state in the coating, such as temperature gradients during and after deposition, stress relaxation processes (plastic deformation, cracking, etc.), phase transformations and chemistry changes [14]. In general, topcoats in TBCs have low residual stresses in as-sprayed condition mainly due to the brittle nature of the ceramic which results in stress relaxation [15].

The stress state in TBCs during service conditions changes as the TGO layer is gradually formed; this phenomenon is discussed in detail in Sect. 3.5.

3.3.2   Young's Modulus

The mechanical properties of thermal sprayed coatings are highly dependent on the microstructure as it is significantly different from that of conventionally processed materials. The elastic modulus or Young's modulus and Poisson's ratio are the basic parameters associated with mechanical behaviour of materials from an engineering context.

Young's modulus is the most commonly used parameter in industry for TBCs to describe their mechanical behaviour. It determines the coating's response under a state of tension or compression. Young's modulus is required to evaluate the parameters describing the material mechanics in TBCs such as thermal stress and residual stress.

Most of the studies on mechanical properties of TBCs have been based on evaluating Young's modulus and elastic anisotropy at low stresses, apart from studies investigating hardness, creep behaviour, etc. It has been observed that ceramics show up to three to ten times lower stiffness constants than the corresponding well-sintered materials [16]. Different Young's moduli in different directions parallel and perpendicular to the surface have also been observed. This anisotropy is attributed to the preferred orientation of the planar defects present in the topcoat, which affects the local compliance substantially and results in a lower value of measured Young's modulus [17]. Analytical models have also been developed which explain this behaviour [16, 18]. Similar behaviour in tension and compression was assumed in these models.

The evaluation of modulus provides indication of the coating integrity, porosity and bonding quality between splats. It also provides an idea of the thermal stress developed in the coating during operation since the modulus is roughly proportional to the induced thermal stresses [19].

3.3.3 Non-linear Properties

Recent developments have shown that ceramic coatings exhibit anelastic mechanical response [20, 21]. Their behaviour both in tension and compression is strongly non-linear. In general, increasing tensile stress results in a lower value of coating modulus [22]. Generally, anelastic responses arise due to two factors: phase transformation and geometrical condition [21]. Since YSZ coatings usually exhibit a stable tetragonal structure during the whole range of operating temperature, phase transformations should not occur. Thus the anelasticity should be due to the geometrical aspects of the coating.

The non-linearity seems to be driven by unique microstructural features present in the TBCs, specifically micro-cracks and weak splat interfaces. The opening and closing of cracks and sliding of sprayed lamellae over each other give rise to a non-linear response. The apparent stiffness decreases with increasing tensile stress as the cracks faces open apart, while it increases with increasing compressive stress as the cracks faces are closed together. The frictional sliding of unbonded interfaces between the splats results in dissipated energy during the loading–unloading cycle thus giving rise to hysteresis.

The anelastic response of a APS YSZ coating during the bilayer curvature measurements using ex situ coating property (ECP) sensor is shown in Fig. 3.5 [2]; the measurement set-up details are discussed further in Sect. 4.3. The coating, initially under a state of compression after deposition, is heated from room temperature to a certain temperature (stress transition from state A to B in Fig. 3.5). Due to the

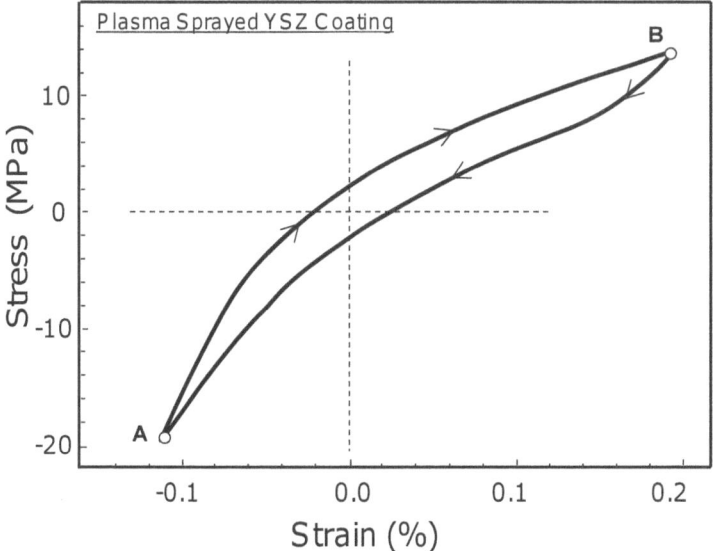

Fig. 3.5 Anelastic response of APS YSZ coating during bilayer curvature measurements [2]

difference in CTE between substrate and coating materials, thermal mismatch stresses arise which result in the change of stress state in the coating from compression to tension during heating and then back to compression as the system is cooled down to the start temperature. The non-linear behaviour of the coating during both heating and cooling part of the cycle can be clearly noticed, as well as the hysteresis in the stress–strain curve. It should be remarked here that the results exhibit nearly complete elastic recovery indicating anelastic response.

3.4 Interface Roughness

3.4.1 Roughness Relationship with Lifetime

Surface roughness is an essential requirement for an APS coating to adhere to the substrate (bondcoat in case of TBCs) on which it is sprayed. Mechanical anchoring is believed to be one of the key mechanisms responsible for adherence of the coating, which would be enhanced by higher surface roughness [23]. If the roughness is too low, the coating adhesion will not be adequate. On the other hand, if the roughness is too high, large thermo-mechanical stresses would be induced on the topcoat, especially due to TGO growth, and the coating could spall off. Additionally, local aluminium depletion could occur near the local protrusions within the bondcoat close to the topcoat–bondcoat interface. This could result in the formation of the fast-growing detrimental spinel oxides and thus even higher stresses. Daroonparvar et al. showed this phenomenon experimentally where Ni infiltrated through the micro-cracks across the Al_2O_3 layer leading to the formation of mixed oxides (CSN) which formed protrusions in the TGO that initiate failure mechanisms of the topcoat [24]. Thus an optimised surface roughness is required for achieving a TBC with long lifetime. Fauchais et al. have shown that the height of surface roughness hills must be about one-third to one half of the mean splat diameter for good adhesion of APS coatings [25]. The ideal level of bondcoat roughness for APS topcoats for good adhesion of the coating is believed to be in the range $Ra = 6–12$ μm [3]. The bondcoat roughness could be influenced by the spray parameters as well the bondcoat powder feedstock size and size distribution [26–28].

Topcoat–bondcoat interface roughness is one of the important parameters which determine the lifetime of a TBC. Eriksson et al. studied this effect by considering four specimens with same chemistry but with different bondcoat roughness. It was observed that increasing Ra resulted in a higher TCF lifetime [29]. These results were in agreement with a detailed modelling work done earlier by Vassen et al. where it was concluded that lower roughness results in longer cracks near the topcoat–bondcoat interface, which would imply earlier failure [28]. However, in another recent work done by Curry et al., two samples with same chemistry as well as similar Ra were found to have significantly different lifetimes, which was understood to be due to the presence of different topographical features [30]. These results show that the traditionally used *Ra* is not sufficient to characterise the coatings and more sophisticated procedures should be used which could characterise the three-dimensional (3D) surface profile in a more precise way.

3.4.2   Stress Inversion Theory

A theory for crack propagation mechanism for APS TBCs has been proposed in earlier works, namely 'stress inversion' theory explained by using a simplified sinusoidal wave profile to represent the topcoat–bondcoat interface [28, 31, 32]. According to the proposed theory, in the initial as-sprayed state without a TGO layer, tensile stresses exist in the hills while compressive stresses exist in the valleys within the topcoat near the topcoat–bondcoat interface as shown in Fig. 3.6a. This stress state is inverted as the TGO layer grows during thermal cyclic loading as shown in Fig. 3.6b. Thus, a crack starts from the hill and propagates to the adjacent valley as the TGO is formed joining the corresponding crack from the other side eventually leading to the spallation of the topcoat as schematically illustrated by the dashed line in Fig. 3.6. The thickness of the TGO when the stress inversion takes place depends on several factors such as the loading conditions, geometry and material [34, 35].

This effect which occurs due to different CTEs of different layers in the TBC could be explained by considering the bondcoat hill as small cylinder surrounded by a concentric cylindrical shell representing the topcoat and topcoat profile as small cylinder surrounded by a concentric cylindrical shell representing the bondcoat for the bondcoat valley [28]. In the as-sprayed state, the metallic bondcoat contracts more due to higher CTE than the ceramic topcoat which leads to tensile stresses in the topcoat near the hills and compressive stresses near the valleys. After the TGO

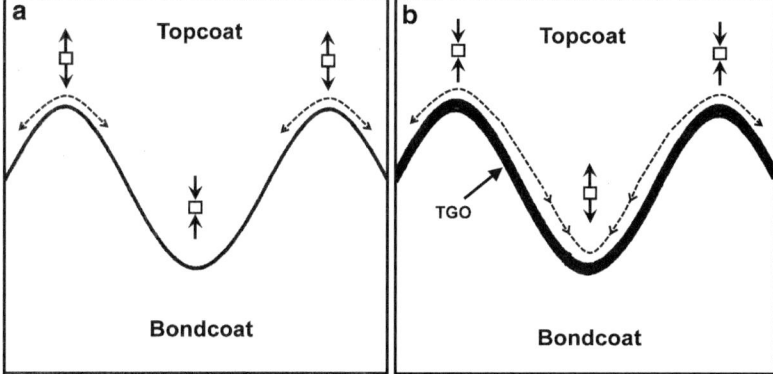

Fig. 3.6 A schematic illustration of the stress behaviour in the topcoat near the bondcoat surface (**a**) in as-sprayed condition and (**b**) after TGO growth [33]

layer forms, the stress state can be understood by replacing the bondcoat by TGO in the concentric cylindrical shell model. Since the TGO is even more stiff (lower CTE) than topcoat, compressive stresses are now introduced in the topcoat near the hills and tensile stresses near the valleys. This explanation has been verified with the help of finite element modelling in several works [32, 36].

The stress inversion theory has been observed to follow the trend in earlier works when the time to stress inversion was compared with the experimental lifetime of TBCs. A 2D sinusoidal profile representing the topcoat–bondcoat interface roughness was used in finite element models by Vassen et al. to evaluate the residual stresses developed in TBCs and it was observed that the time to stress inversion was shorter for samples which failed earlier in experiments [28].

3.5 Oxide Formation

As soon as the TBC is put into operating conditions, the bondcoat starts to undergo oxidation due to the exposure to high temperatures. The YSZ topcoat used typically in a TBC is transparent to oxygen due to two effects: (1) zirconia is transparent to oxygen flow due to the presence of vacancies and (2) the interconnected porosity network present within the topcoat allows free flow of oxygen (air). Therefore, the bondcoat metallic alloy is designed to act as a local aluminium reservoir allowing the formation of slow growing α-alumina which could provide oxidation resistance to the substrate. Alumina is the primary and most stable oxide formed for a NiCoCrAlY bondcoat during operation [37]. This is due to the fact that alumina has a lower formation energy than the other main element such as Ni, Co and Cr present in the bondcoat [38]. The formation energy of alumina ($\Delta G°$) is given by the following relationship:

$$\Delta G° = RT \ \ln\left(a_{Al}^{4/3} \cdot P_{O2}\right) \tag{3.6}$$

where R is the gas constant, T is the temperature in Kelvin, a_{Al} is the activity of aluminium and P_{O2} is the partial pressure of oxygen. The activity of aluminium depends on the bondcoat composition.

The alumina layer is supposed to suppress the formation of other detrimental oxides during the extended thermal exposure in service thus improving lifetime of the TBC. However, other oxides such as CSN are formed at a later stage due to microstructural defects such as cracks present within the alumina layer and depletion of alumina within the bondcoat depending on several factors such as the composition of the bondcoat and the operating conditions as described by (3.6).

The oxide formation process at high temperatures typically during TCF testing is usually categorised in three stages [38, 39]:

1. **Transient stage**: In this stage, oxides form at a rapid rate. Almost all of the alloying element present in the bondcoat can be oxidised during this stage, the variety mainly dependent on the chemical composition and microstructure of

the coating. This stage is typically very short, in the range of less than one hour for Ni–Cr–Al alloys above 1,000 °C [40].

2. **Steady stage**: In this stage, alumina grows in a continuous and dense form linearly which acts as a protective layer. The alumina layer could be cracked or spalled off due to thermo-mechanical stresses and reformed [38].

3. **Acceleration stage**: When the aluminium content in the bondcoat becomes lower than a critical level, other mixed oxides such as CSN start to form aggressively leading eventually to failure.

In case of alumina, the oxide growth is believed to be mainly driven due to two parallel diffusion mechanisms: inward diffusion of oxygen, mostly along grain boundaries, towards the TGO–bondcoat interface and outward diffusion of aluminium along the already formed alumina particles towards the topcoat–TGO interface [37]. However, it is generally accepted that the inward growth of alumina is the predominant step and the reaction occurs primarily at the bondcoat–TGO interface [41, 42].

The growth of the TGO layer causes stress in the TBC both due to swelling of TGO resulting in 'growth stresses' as well as stresses generated due to the mismatch in CTEs between the different layers in the TBC. The growth stresses are usually relaxed quite quickly due to the high creep rates in the ceramic topcoat layer at elevated temperatures [34]. However, the thermal mismatch stresses, generated especially near the topcoat–bondcoat interface when the TBC is cooled, could be detrimental and result in the formation of cracks and/or elongation of pre-existing cracks near the topcoat–bondcoat interface which could eventually lead to failure [34].

The growth of mixed oxides is highly disadvantageous for a TBC as these oxides grow at a rapid rate and have an associated volume expansion which could result in very high growth stresses leading to spallation and thus failure. It has been shown that these oxides form protrusions in the TGO which could initiate failure mechanisms of the topcoat [24]. Bulky mixed oxide clusters may form from individual splats which are cut off from the rest of the bondcoat and quickly oxidise as they run out of aluminium. It has been observed in previous work that these oxide clusters are prone to cracking and CSN-nucleated cracks would assist the crack coalescence process leading to the formation of large cracks that could result in spallation and thus failure [39]. Eriksson et al. observed a significant reduction in TCF lifetime of TBCs which formed more CSN [29].

3.6   Failure Mechanisms

Failure mechanisms in TBCs are very complex and are usually a mixture of several mechanisms. The primary failure method is spallation of the coating due to a fracture in the ceramic layer and/or the TGO layer. These types of failures occur if either the local stresses increase, the material strength decreases or a combination of the two. The stress can increase due to mismatch in CTE, temperature gradient and TGO formation. The changes in stress distribution induce the propagation of pre-existing cracks near the interface finally leading to cracks long enough to cause spallation of the topcoat. The formation of mixed oxides can significantly enhance the crack propagation and coalescence as discussed in detail in Sect. 3.5. The topcoat could become more brittle due to sintering at high temperatures and/or propagation of cracks, while the bondcoat could become more brittle due to depletion of aluminium which would lead to growth of more brittle oxides.

Some of the other failure mechanisms of a TBC are: (1) damage induced by particle impact, such as erosion and foreign object damage (FOD), (2) cracks formed due to the penetration of deposits of calcium–magnesium–aluminosilicate (CMAS) formed due to ingress of sand and dust present in the atmosphere into the turbine engine [43, 44].

References

1. Curry N, Donoghue J (2012) Evolution of thermal conductivity of dysprosia stabilised thermal barrier coating systems during heat treatment. Surf Coat Technol 2019:38–43
2. Dwivedi G (2011) On the anelastic behavior of plasma sprayed ceramic coatings: observations, characterizations and applications. Ph.D. thesis, Stony Brook University, USA
3. Curry N (2014) Design of thermal barrier coatings. Ph.D. thesis, University West, Sweden, No. 3.
4. Friis M (2002) A methodology to control the microstructure of plasma sprayed coatings. Ph.D. thesis, Lund University, Sweden
5. Hasselman DPH (1978) Effect of cracks on thermal conductivity. J Compos Mater 12:403–407
6. Golosnoy IO, Cipitria A, Clyne TW (2009) Heat transfer through plasma-sprayed thermal barrier coatings in Gas turbines: a review of recent work. J Therm Spray Technol 18(5–6): 809–821
7. Klemens PG (1969) Theory of thermal conductivity in solids. In: Thermal conductivity, vol. 1, Tye RP (ed.) Academic, London, pp 1–68
8. Pawlowski L, Fauchais P (1992) Thermal transport properties of thermally sprayed coatings. Int Mater Rev 37(6):271–289
9. Klemens PG, Gell M (1998) Thermal conductivity of thermal barrier coatings. Mater Sci Eng A 245(2):143–149
10. Nicholls JR, Lawsona KJ, Johnstone A, Rickerby DS (2002) Methods to reduce the thermal conductivity of EB-PVD TBCs. Surf Coat Technol 151–152:383–391
11. Golosnoy IO, Tripas SA, Clyne TW (2005) An analytical model for simulation of heat flow in plasma-sprayed thermal barrier coatings. J Therm Spray Technol 14(2):205–214
12. Lim G, Kar A (2009) Modeling of thermal barrier coating temperature due to transmissive radiative heating. J Mater Sci 44:3589–3599
13. Clyne TW, Gill SC (1996) Residual stresses in thermal spray coatings and their effect on interfacial adhesion: a review of recent work. J Therm Spray Technol 5(4):401–418
14. Matejicek J, Sampath S, Dubsky J (1998) X-ray residual stress measurement in metallic and ceramic plasma sprayed coatings. J Therm Spray Technol 7(4):489–496
15. Matejicek J, Sampath S, Brand PC, Prask HJ (1999) Quenching, thermal and residual stress in plasma sprayed deposits: NiCrAlY and YSZ coatings. Acta Mater 47(2):607–617
16. Kroupa F (2007) Nonlinear behavior in compression and tension of thermally sprayed ceramic coatings. J Therm Spray Technol 16(1):84–95
17. Guo S, Kagawa Y (2004) Young's moduli of zirconia top-coat and thermally grown oxide in a plasma-sprayed thermal barrier coating system. Scr Mater 50:1401–1406

18. Sevostianov I, Kachanov M (2001) Plasma-sprayed ceramic coatings: anisotropic elastic and conductive properties in relation to the microstructure; cross-property correlations. Mater Sci Eng A 297(1–2):235–243
19. Matejicek J, Sampath S (2003) In situ measurement of residual stresses and elastic moduli in thermal sprayed coatings Part 1: apparatus and analysis. Acta Mater 51:863–872
20. Liu Y, Nakamura T, Srinivasan V, Vaidya A, Gouldstone A, Sampath S (2007) Non-linear elastic properties of plasma-sprayed zirconia coatings and associated relationships with processing conditions. Acta Mater 55(14):4667–4678
21. Liu YJ, Nakamura T, Dwivedi G, Valarezo A, Sampath S (2008) Anelastic behavior of plasma-sprayed Zirconia coatings. J Am Ceram Soc 91(12):4036–4043
22. Nakamura T, Liu YJ (2007) Determination of nonlinear properties of thermal sprayed ceramic coatings via inverse analysis. Int J Solids Struct 44(6):1990–2009
23. Nusair Khan A, Lu J, Liao H (2003) Effect of residual stresses on air plasma sprayed thermal barrier coatings. Surf Coat Technol 168:291–299
24. Daroonparvar M, Hussain MS, Yajid MAM (2012) The role of formation of continues thermally grown oxide layer on the nanostructured NiCrAlY bond coat during thermal exposure in air. Appl Surf Sci 261:287–297
25. Fauchais P, Etchart-Salas R, Rat V, Coudert JF, Caron N, Wittmann-Teneze K (2008) Parameters controlling liquid plasma spraying solutions, sols, or suspensions. J Therm Spray Technol 17(1):31–59
26. Nowak W, Naumenko D, Mor G, Mor F, Mack DE, Vassen R, Singheiser L, Quadakkers WJ (2014) Effect of processing parameters on MCrAlY bondcoat roughness and lifetime of APS–TBC systems. Surf Coat Technol 260:82–89
27. Rajasekaran B, Mauer G, Vaßen R (2011) Enhanced characteristics of HVOF-sprayed MCrAlY bond coats for TBC applications. J Therm Spray Technol 20(6):1209–1216
28. Vaßen R, Kerkhoff G, Stöver D (2001) Development of a micromechanical life prediction model for plasma sprayed thermal barrier coatings. Mater Sci Eng A 303:100–109
29. Eriksson R, Sjöström S, Brodin H, Johansson S, Östergren L, Li X-H (2013) TBC bond coat–top coat interface roughness: Influence on fatigue life and modelling aspects. Surf Coat Technol 236:230–238
30. Curry N, Markocsan N, Östergren L, Li X-H, Dorfman M (2013) Evaluation of the lifetime and thermal conductivity of dysprosia-stabilized thermal barrier coating systems. J Therm Spray Technol 22(6):864–872
31. Pindera M-J, Aboudi J, Arnold SM (2000) The effect of interface roughness and oxide film thickness on the inelastic response of thermal barrier coatings to thermal cycling. Mater Sci Eng A 284:158–175
32. Jinnestrand M, Sjöström S (2001) Investigation by 3D FE simulations of delamination crack initiation in TBC caused by alumina growth. Surf Coat Technol 135:188–195
33. Gupta M, Eriksson R, Sand U, Nylén P (2014) A diffusion-based oxide layer growth model using real interface roughness in thermal barrier coatings for lifetime assessment. Surf Coat Technol. doi 10.1016/j.surfcoat.2014.12.043
34. Vaßen R, Giesen S, Stöver D (2009) Lifetime of plasma-sprayed thermal barrier coatings: comparison of numerical and experimental results. J Therm Spray Technol 18(5–6):835–845
35. Bäker M (2012) Finite element simulation of interface cracks in thermal barrier coatings. Comput Mater Sci 64:79–93
36. Ahrens M, Vaßen R, Stöver D (2002) Stress distributions in plasma-sprayed thermal barrier coatings as a function of interface roughness and oxide scale thickness. Surf Coat Technol 161:26–35
37. Busso EP, Lin J, Sakurai S, Nakayama M (2001) A mechanistic study of oxidation-induced degradation in a plasma-sprayed thermal barrier coating system. Part I: model formulation. Acta Mater 49:1515–1528
38. Yuan K (2014) Oxidation and corrosion of new MCrAlX coatings: modelling and experiments. Ph.D. thesis, Linköping University, Sweden

39. Chen WR, Wu X, Marple BR, Patnaik PC (2006) The growth and influence of thermally grown oxide in a thermal barrier coating. Surf Coat Technol 201:1074–1079
40. Haynes JA, Rigney ED, Ferber MK, Porter WD (1996) Oxidation and degradation of a plasma-sprayed thermal barrier coating system. Surf Coat Technol 86–87:102–108
41. Hille TS, Turteltaub S, Suiker ASJ (2011) Oxide growth and damage evolution in thermal barrier coatings. Eng Fract Mech 78:2139–2152
42. Ma K, Schoenung JM (2011) Isothermal oxidation behavior of cryomilled NiCrAlY bond coat: homogeneity and growth rate of TGO. Surf Coat Technol 205:5178–5185
43. Vaßen R, Jarligo MO, Steinke T, Mack DE, Stöver D (2010) Overview on advanced thermal barrier coatings. Surf Coat Technol 205:938–942
44. Drexler JM, Chen C-H, Gledhill AD, Shinoda K, Sampath S, Padture NP (2012) Plasma sprayed gadolinium zirconate thermal barrier coatings that are resistant to damage by molten Ca–Mg–Al–silicate glass. Surf Coat Technol 206:3911–3916

Chapter 4
Experimental Methods

4.1 Microstructure Characterisation

Various techniques have been used in the past for qualitative and/or quantitative analysis of microstructure of TBCs. Mercury intrusion porosimetry (MIP) enables the measurement of total porosity for open pores and the evaluation of pore size distribution over a wide range [1, 2]. Small-angle neutron scattering (SANS) can obtain statistical results for small scale defects as well as orientation information [3–5]. Image analysis is a robust, reliable and inexpensive method to characterise TBC microstructures from cross-section images which can be used to obtain information about porosity, pores and cracks distribution and orientation, etc. [6]. The commonly used techniques to capture microstructure cross-section images of TBCs are LOM and SEM.

During the image analysis procedure, several images are taken along the coating cross section to capture the variation in the microstructure. These images are then converted to binary format by thresholding using an image analysis software such as Aphelion (ADCIS, France) and Image J (National Institutes of Health, USA). The images have to be taken carefully at a certain magnification so as to capture the relevant coating microstructure details. High magnification images result in a loss of global coating information, while too low magnification images are unable to capture small scale details present in the microstructure. Other factors which could affect the image analysis results are threshold level, microscope effects (such as image brightness and contrast and operating mode) and coating location where the image has been taken [6].

4.2 Thermal Conductivity Measurements

The most widely accepted method for evaluating thermal conductivity of TBCs is laser flash analysis (LFA) technique [7]. The state-of-the-art measurement set-up is shown in Fig. 4.1.

© The Author(s) 2015
M. Gupta, *Design of Thermal Barrier Coatings*, SpringerBriefs in Materials,
DOI 10.1007/978-3-319-17254-5_4

Fig. 4.1 The laser flash analysis apparatus used for measuring thermal conductivity (*Courtesy of NETZSCH, Germany*)

In this method, a laser pulse is shot at the substrate face of the sample and the resulting temperature increase is measured at the other face of the sample with an infrared detector. The signal is then normalised and the thermal diffusivity is calculated using an equation based on one of the existing models. One such equation is as follows:

$$\alpha = \left(0.1388 L^2\right) / t_{(0.5)} \tag{4.1}$$

where α is the thermal diffusivity, L is the thickness of the sample and $t_{(0.5)}$ is the time taken for the rear face of the sample to reach half of its maximum rise. Thermal conductivity (K) can then be calculated if the density (ρ) and specific heat capacity (C_p) of the coating are known using the following relation:

$$K = \alpha.C_p.\rho \tag{4.2}$$

Samples are normally coated with a thin layer of graphite or gold before the measurements. Since zirconia is translucent to light in the wavelength of the laser, the presence of graphite layer is essential to prevent the laser pulse from travelling through the ceramic layer.

Density can be measured using one of the several existing methods such as Archimedes displacement, mercury porosimetry and gas absorption methods. These measurements are based on the assumption that all porous features present in the coating are interconnected, which is not true in every case. An indirect approach for evaluating density could be to measure the porosity level from image analysis of microstructure cross-section images and then calculating the density based on the density of bulk material. Usually, the density evaluated using image analysis method is lower than the mercury porosimetry method, which could be attributed to the effect of unconnected or sealed porosity detected by the image analysis method. Specific heat capacity data of the coatings can be used from existing databases, though it is recommended to measure it for each coating material using differential scanning calorimetry (DSC) for accuracy.

Thermal conductivity is measured either on free-standing ceramic coating, which can be peeled off from the substrate, or on the complete TBC system. In the latter case, thermal conductivity of topcoat is calculated afterwards by measuring the thicknesses of individual layers and thermal conductivity values of only the substrate and substrate sprayed with bondcoat. Thicknesses are measured from microstructure cross-section images using an optical microscope. The measurement of thickness can be difficult due to high roughness of the interfaces in TBC systems and can result in high errors in final measurements since thermal diffusivity is proportional to the square of thickness [7].

LFA technique is essentially non-contact and the equipment is available commercially from several companies. A major drawback is that only small flat samples can be used for measurement and thus measurement on a real component is not possible.

4.3   Young's Modulus Measurements

Traditional techniques used for measuring Young's modulus of APS topcoats are micro-indentation, four-point bending, etc. Another technique which has been developed in recent years is bilayer curvature measurements. The ECP sensor which consists of non-contact displacement lasers and thermocouples to monitor the displacement and temperature simultaneously at the back of the sample during a heating and cooling cycle is used for these measurements [8]. The bilayer curvature measurement set-up is shown in Fig. 4.2.

The techniques used to measure Young's modulus of TBCs can be classified into three categories [9]:

1. *Mechanical loading methods*: These techniques are based on the direct measurement of the deformation of the test material upon application of an external force. Examples include bending tests of coating system or a freestanding coating, tensile or compression tests and indentation tests.

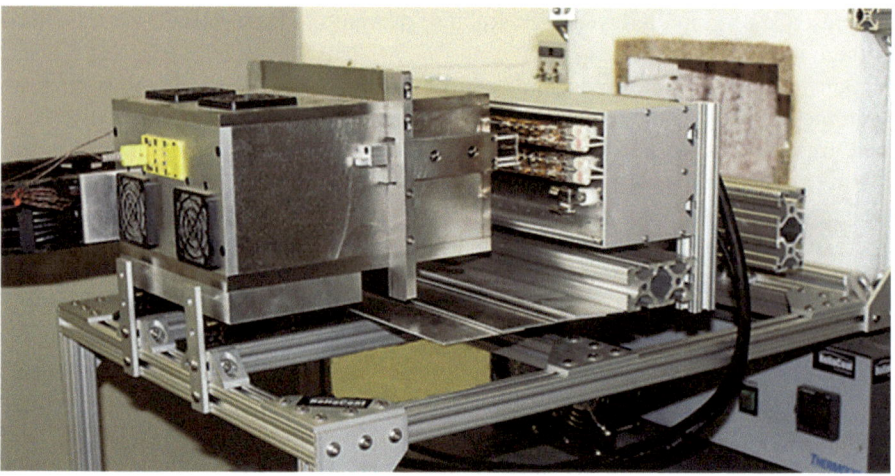

Fig. 4.2 The bilayer curvature measurement set-up using the ex situ coating property sensor (*Courtesy of ReliaCoat Technologies, NY, USA*)

2. *Resonance vibration methods*: These techniques are based on the principle that the resonance frequency of a material depends on its elastic modulus of the material. Examples include acoustic emission techniques [10].
3. *Pulse-echo methods*: These techniques are based on the principle that the velocity of sound in a material depends on the elastic modulus of the material. Examples include laser-induced ultrasonic techniques [11].

It has been observed in earlier work that the modulus value depends highly on the technique used to evaluate it and it could be different even if the test specimens are identical between different techniques [12]. This effect, which is related to the local variations of the microstructure within the coating, could be explained by the scale of the testing technique which ranges from microscopic (nanoindentation) to mesoscopic (micro-indentation) and macroscopic (bending tests, ultrasonic testing) dimensions [12]. Therefore, elastic modulus needs to be considered as an engineering value for quantifying and ranking coating systems rather than as a fundamental property.

4.4   Roughness Measurements

The traditional technique to measure surface roughness is to use a surface profilometer. Surface profilometer is a contact technique which determines the profile of a surface by dragging a stylus along it. The equipment is able to output the 2D roughness parameters such as *Ra*, *Rq* and *Rz*. This technique is cheap and very simple to use.

However, recent works have shown the 2D surface roughness parameters might not be sufficient to characterise a coating [13]. Other techniques such as white light

Table 4.1 Some of the ISO 25178 feature parameters calculated for HVOF and APS bondcoat samples

Parameter (units)	Parameter description
Spd (1/mm^2)	Density of hills (number of hills per unit area)
Spc (1/mm)	Arithmetic mean hill curvature (arithmetic mean of the principal curvatures of hills within a definition area)
Sha (mm^2)	Mean hill area (average area of the hills connected to the edge at a particular height)

interferometry and laser stripe projection could be useful to analyse these coatings. The major advantages of using these techniques are: (1) A 3D profile could be captured with these techniques which could provide better visualisation of the surface features. (2) Very low resolution can be obtained with these techniques; for example, the vertical resolution obtained with interferometry is in the range of nanometres. (3) The newly formulated ISO 25178 parameters could be measured and calculated with the help of these techniques thus enabling to characterise the 3D profile quantitatively. The disadvantages of using these techniques are that they are very expensive and the data analysis could be a complex procedure.

The feature parameters are a new family of parameters which has been integrated into the ISO 25178 standard. Feature parameters are derived by creating segmentation motifs over a surface which makes it possible to identify specific areas such as hills and valleys. Some of the ISO feature parameters are given in Table 4.1. Quantifying the density, size and shape of the hills and valleys of the bondcoat surface could be highly beneficial to assess the stresses' characteristics near the topcoat–bondcoat interface.

Figure 4.3a shows a surface profile of a bondcoat sample captured using white light interferometry. The figure shows the presence of large hills over the surface which are suspected to be partially melted/unmelted particles formed during the spray process. Figure 4.3b shows segmentation motifs captured using stripe projection technique for the same sample. The presence of the hills over the surface could be quantitatively analysed with the help of this image.

4.5 Lifetime Testing

Assessing the lifetime of TBCs is a challenging task since testing them in real environments is very time consuming and expensive. Accelerated and simplified tests are used to reduce testing times by exposing the TBCs to harsher environments than reality. Due to the over-simplification of testing methods, there can be large discrepancies between the performance in testing and operating conditions. Nevertheless, the lifetime tests provide a fair idea of the coating behaviour and can be used to judge different coatings. Common methods for testing the lifetime of TBCs are TCF and thermal shock tests. Both these methods are well practised in the industry.

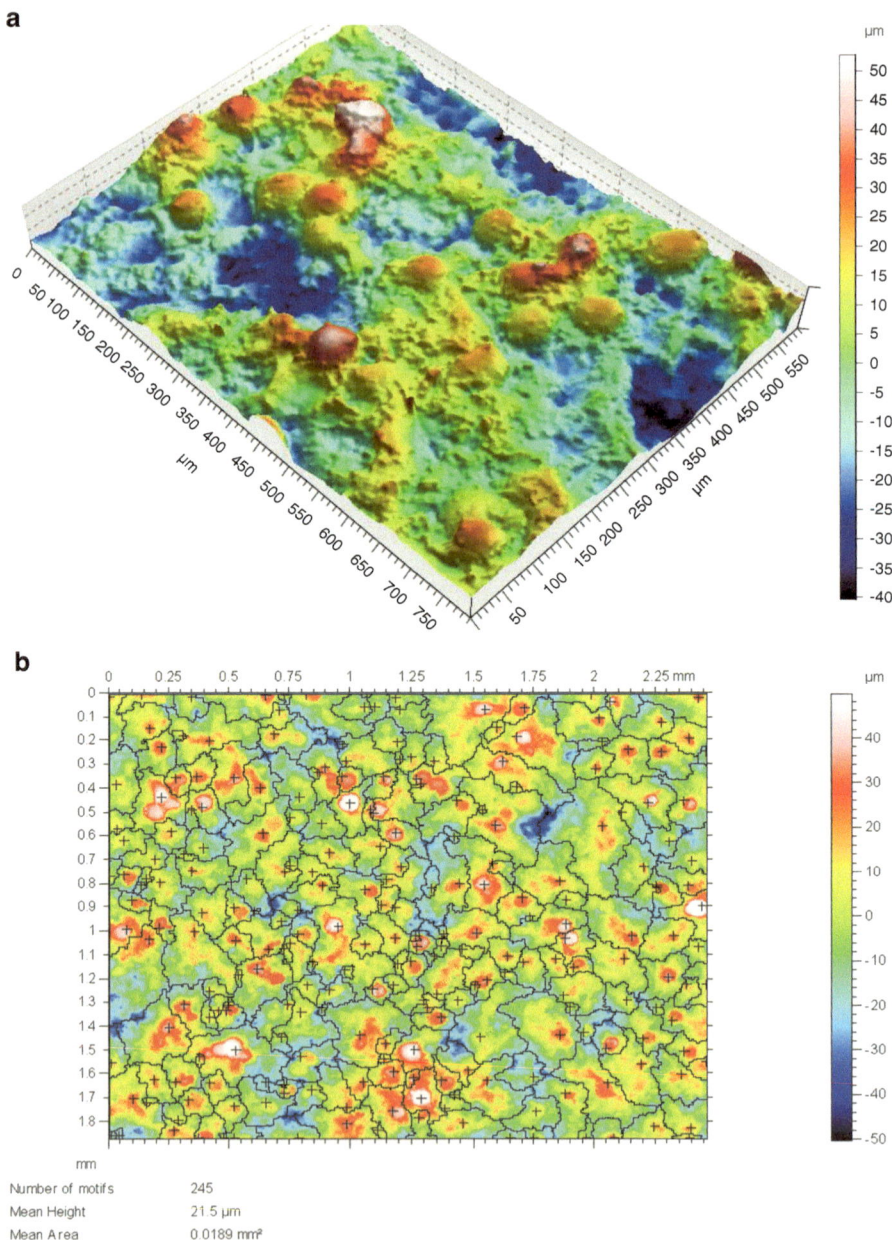

Fig. 4.3 (**a**) Bondcoat surface profiles captured using white light interferometry technique and (**b**) segmentation motifs captured using stripe projection technique [14]

In TCF tests, the samples are typically first heated in a furnace at high temperatures around 1,100 °C for one hour and then cooled down for ten minutes using compressed air to approximately 100 °C. These two steps are repeated until failure. The criterion for failure is deemed to be more than 20 % visible spallation of the coating.

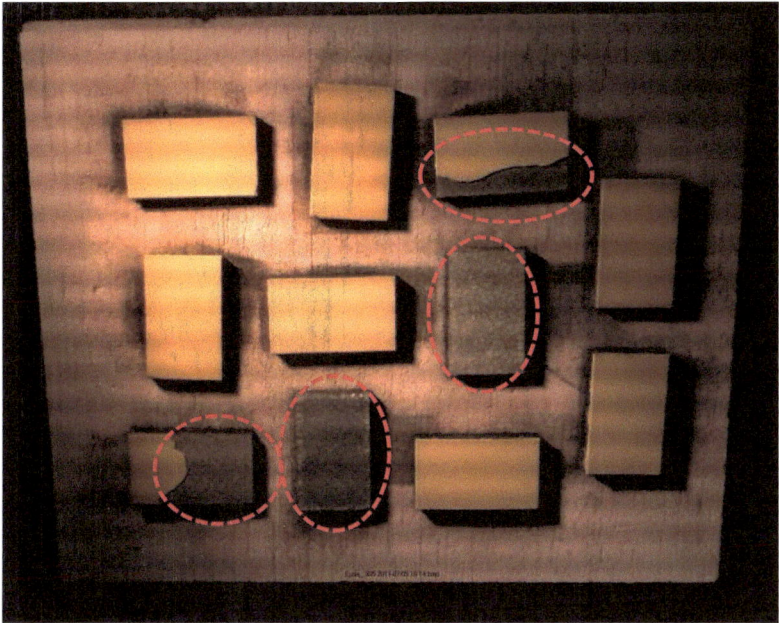

Fig. 4.4 Samples during TCF testing showing spallation of a few samples; the indicated areas show the failed parts of the samples

A visual record of the samples' surface is made at the end of each cycle using a webcam which is used later for calculating the number of cycles to failure. Figure 4.4 shows one such photograph taken during the testing where it can be seen that a few of the samples have the topcoats spalled off while a few are still intact. The failed areas in the samples are indicated in the figure.

Thermal shock testing simulates the rapid heating and cooling experienced by coatings during service. As the name suggests, the coatings are heated up to a specified temperature and then cooled down rapidly with air or by dipping them into water. Burner rig test is a common way of testing thermal shock behaviour. In this test, samples are heated on the coating face with a combustion burner flame and then cooled down with compressed air. Samples are cooled on the back face with water or air to set up a thermal gradient within the TBC system. A typical cycling time for burner rig tests is five minutes heating and two minutes cooling.

Figure 4.5 shows a microstructure cross-section image of part of the topcoat–bondcoat interface after failure during TCF test. The topcoat zirconia layer was sprayed by APS while the bondcoat NiCoCrAlY layer was sprayed by HVOF technique. The dark grey layer in the image between the topcoat and bondcoat consists of TGO formed during the test.

Both TCF and thermal shock testing methods have their own limitations. TCF tests have the benefit that they provide information about sintering of topcoat, TGO growth in TBC system and the topcoat's ability to sustain the induced stresses due to TGO growth and thermal mismatch. In this way, results obtained from TCF tests

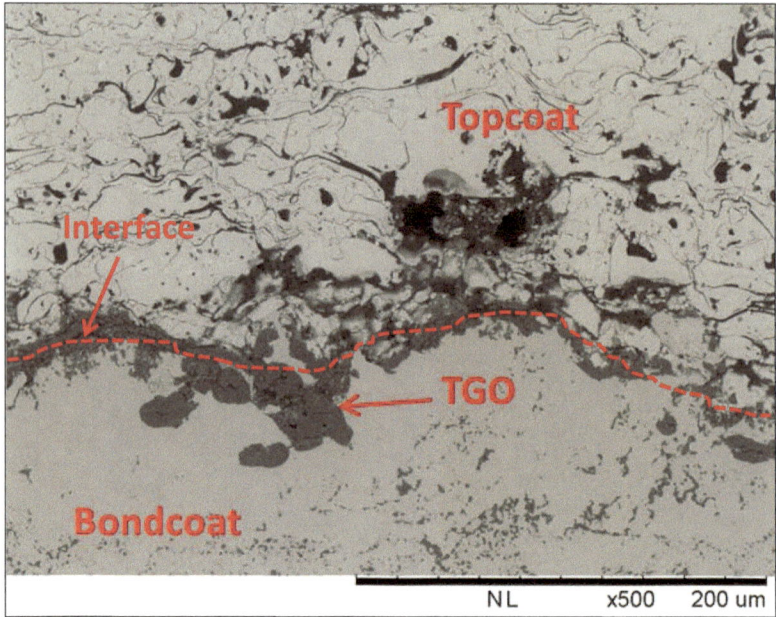

Fig. 4.5 Topcoat–bondcoat interface after failure showing the TGO growth

are highly dependent on the bondcoat material and structure. The major limitation of TCF tests is the time taken for the tests to be completed, since it can take more than a month to obtain lifetime results. Another limitation is that since the samples are normally maintained in an isothermal heating environment, a thermal gradient is not present in the TBC system which is present in service conditions. Homogeneous temperature state can lead to higher stresses and thus lower lifetime than reality [15]. Thermal shock tests induce large thermal stresses in the topcoat–bondcoat interface and they are mainly dependent on the strain tolerance of topcoats. There is little time in these tests for any significant effect of bondcoat oxidation or sintering of topcoats on coating lifetime. Though thermal shock tests are much faster as they normally take less than a week to be completed, they have a disadvantage that they are much more severe than reality. An alternative testing method which involves a combination of TCF and thermal shock tests was suggested in an earlier work for better representation of actual engine conditions [16].

References

1. Portinhaa A, Teixeira V, Carneiro J, Martins J, Costa MF, Vassen R, Stoever D (2005) Surf Coat Technol 195:245–251
2. Cernuschi F, Golosnoy IO, Bison P, Moscatelli A, Vassen R, Bossmann H-P, Capelli S (2013) Microstructural characterization of porous thermal barrier coatings by IR gas porosimetry and sintering forecasts. Acta Mater 61(1):248–262

3. Kulkarni A, Wang Z, Nakamura T, Sampath S, Goland A, Herman H, Allen J, Ilavsky J, Long G, Frahm J, Steinbrech RW (2003) Comprehensive microstructural characterization and predictive property modeling of plasma-sprayed zirconia coatings. Acta Mater 51(9):2457–2475

4. Wang Z, Kulkarni A, Deshpande S, Nakamura T, Herman H (2003) Effects of pores and interfaces on effective properties of plasma sprayed zirconia coatings. Acta Mater 51(18): 5319–5334

5. Allen AJ, Ilavsky J, Long GG, Wallace JS, Berndt CC, Herman H (2001) Microstructural characterization of yttria-stabilized zirconia plasma-sprayed deposits using multiple small-angle neutron scattering. Acta Mater 49(9):1661–1675

6. Tan Y (2007) Thermal design and microstructure-based property assessment for thermal spray coating systems, Ph.D. thesis, Stony Brook University, USA

7. Taylor RE (1998) Thermal conductivity determinations of thermal barrier coatings. Mater Sci Eng A 245(2):160–167

8. Matejicek J, Sampath S, Brand PC, Prask HJ (1999) Quenching, thermal and residual stress in plasma sprayed deposits: NiCrAlY and YSZ coatings. Acta Mater 47(2):607–617

9. Waki H, Oikawa A, Kato M, Takahashi S, Kojima Y, Ono F (2014) Evaluation of the accuracy of Young's moduli of thermal barrier coatings determined on the basis of composite beam theory. J Therm Spray Technol 23(8):1291–1301

10. Ma XQ, Mizutani Y, Takemoto M (2001) Laser-induced surface acoustic waves for evaluation of elastic stiffness of plasma sprayed materials. J Mater Sci 36:5633–5641

11. Patsias S, Tassini N, Lambrinou K (2006) Ceramic coatings: effect of deposition method on damping and modulus of elasticity for yttria-stabilized zirconia. Mater Sci Eng A 442: 504–508

12. Margadant N, Neuenschwander J, Stauss S, Kaps H, Kulkarni A, Matejicek J, Rössler G (2006) Impact of probing volume from different mechanical measurement methods on elastic properties of thermally sprayed Ni-based coatings on a mesoscopic scale. Surf Coat Technol 200(8):2805–2820

13. Curry N, Markocsan N, Östergren L, Li X-H, Dorfman M (2013) Evaluation of the lifetime and thermal conductivity of dysprosia-stabilized thermal barrier coating systems. J Therm Spray Technol 22(6):864–872

14. Gupta M, Skogsberg K, Nylén P (2014) Influence of topcoat-bondcoat interface roughness on stresses and lifetime in thermal barrier coatings. J Therm Spray Technol 23(1–2):170–181

15. Ranjbar-Far M, Absi J, Shahidi S, Mariaux G (2011) Impact of the non-homogenous temperature distribution and the coatings process modeling on the thermal barrier coatings system. Mater Des 32:728–735

16. Bolcavage A, Feuerstein A, Foster J, Moore P (2004) Thermal shock testing of thermal barrier coating/bondcoat systems. J Mater Eng Perform 13(4):389–397

Chapter 5
Modelling of Properties of TBCs

Since the macroscopic properties of TBCs are highly dependent on their microscopic structure, modelling of properties of TBCs usually consists of a representation of the present porosity dispersed within a solid phase. Several studies have been performed in the past using both analytical and numerical models for evaluating thermal–mechanical properties of TBCs; a few of them are discussed in Sects. 5.1 and 5.2. The approach used by the author in a recent work is described in Sects. 5.3 and 5.4, and the results using this approach in a recent work are presented in Sect. 5.5 [1–4].

5.1 Thermal Conductivity

5.1.1 Analytical Models

Since the end of nineteenth century, several analytical models have been developed for determining the thermal conductivity of multiphase solids, especially in porous materials. Brief reviews of such works have been done earlier [5–10]. Most of these models consisted of randomly distributed non-interacting pores or ellipsoids, or periodic structures. The first model developed by Maxwell–Eucken is given as [5]

$$\frac{\kappa_{tot}}{\kappa_d} = \frac{1-P}{1+0.5P} \tag{5.1}$$

where K_{tot} is the thermal conductivity of a porous material, K_d is the thermal conductivity of a dense material and P is the fraction of porosity, assuming that the thermal conductivity of pores K_p is small compared to K_d. More recently, another model was developed by Klemens which is given as [11]

$$\frac{\kappa_{tot}}{\kappa_d} = 1 - \frac{4}{3}P \tag{5.2}$$

© The Author(s) 2015
M. Gupta, *Design of Thermal Barrier Coatings*, SpringerBriefs in Materials,
DOI 10.1007/978-3-319-17254-5_5

Hasselman [12] developed a model which calculated the effect of cracks of various orientations on the thermal conductivity of solid materials. It was concluded that maximum thermal insulation is obtained with cracks perpendicular to the direction of heat flow while cracks parallel to the direction of heat flow have no effect on thermal conductivity.

In this book, focus is placed on models based on plasma-sprayed coatings.

The first analytical model for predicting thermal conductivity of plasma-sprayed coatings was proposed by McPherson [13] which involved regions of good and poor contact between lamellae where the regions of poor contact act as thermal resistances. The model can be given as [13]

$$\frac{\kappa_{tot}}{\kappa_d} = \frac{2f\delta}{\pi a} \tag{5.3}$$

where f is the fraction of 'true contact', δ is the lamellae thickness and a is the radius of individual contact areas. Li et al. [14] further developed this model and proposed quantitative structural parameters to characterise the deposit lamellar structure instead of porosity content, the most important parameter being the bonding ratio at the interfaces between lamellae. Thus, they incorporated the thermal resistance of the lamellae into the model. Boire-Lavigne et al. [15] also included the oxide layer resistance in the contact areas in the model. They used this analogy to determine the thermal diffusivity of plasma-sprayed tungsten based coatings, where the geometrical parameters were calculated from an image analysis procedure.

Bjorneklett et al. [16] compared experimental values of thermal diffusivity for ZrO_2-7 % Y_2O_3 and Al_2O_3-3 % TiO_2 plasma-sprayed coatings with analytical models based on effective medium theories. Models representing three different microstructures were considered—a continuous ceramic with dispersed pores, a continuous ceramic matrix with continuously interconnected pores and dispersed ceramic loosely bonded together. It was shown in this work that last two microstructure models conformed better to the theoretical values compared to the first one. This implied that the thermal conductivity model consisting of an agglomerate of ceramic particles behaves similar to reality compared to a solid ceramic sheet with pores. This behaviour is also in agreement with the formation process of a plasma-sprayed coating.

Sevostianov et al. [17] developed a model which gave an explicit relation between the anisotropic thermal conductivities and the microstructure parameters of plasma-sprayed TBCs. The microstructure was assumed to be composed of two families of penny-shaped cracks—horizontal and vertical. The orientational scatter was also accounted for in the model by an appropriate orientation distribution function. The effective conductivity was dependent mainly on the crack densities and their orientational scatters, and the overall porosity (P) played only a secondary role [17]. The thermal conductivity in the direction perpendicular to the substrate was given by the expression [17]

$$\frac{\kappa_{tot}}{\kappa_d} = 1 - \frac{8\alpha_v}{3(1-P)} \tag{5.4}$$

where α_v is the component of crack density tensor in the direction perpendicular to the substrate that incorporates the crack densities of both pore families and their orientational scatters. An explicit correlation between elastic and conductive properties, dependent only on the elastic and conductive properties of the bulk material, was also derived later (Ref [45]). The simplified expression is given as [18]

$$\frac{E_d - E_{tot}}{E_{tot}} \approx 2.14 \frac{\kappa_{tot} - \kappa_d}{\kappa_d} \tag{5.5}$$

where E_d is Young's modulus of bulk material and E_{tot} is Young's modulus in the direction normal to the substrate. More accurate correlations for the two cases of ideally parallel and randomly oriented cracks were also given later, which indicated that the actual cross-property factor would be slightly lower than the one in the last equation [19]. Equations (5.4) and (5.5) were also verified on YSZ coatings by using the image analysis data from photomicrographs of the coatings and comparing the predicted values with experimental values which were in good agreement [19].

Lu et al. [20] made a model which included different kinds of pore morphologies—randomly oriented pores, aligned but spatially random pores, periodic pores and zigzag pores, with distributed pore shapes. Finite element method (FEM) was used to compute the conductivity related to zigzag pores. They applied the model on TBCs made by EB-PVD.

Golosnoy et al. [6] developed an analytical model which comprised solid layers separated by thin, periodically bridged, gas-filled voids. The model was based on dividing the material into two independent zones—one characterised by unidirectional serial heat flow occurring through lamellae and pores and the other by channelled conduction through the bridges between the lamellae. Radiative heat transfer and the effect of gas pressure inside the pores were also included in the model, and thermal conductivity was computed over a range of temperatures. Cipitria et al. [21] used this geometrical model to develop a sintering model based on the application of the variational principle to diffusional phenomena. It also accounted for the effects of surface diffusion, grain boundary diffusion and grain growth. Good agreement was found between these two models and experimental results.

5.1.2 Numerical Models

Hollis [22] developed a model using FEM to compute the thermal conductivity of VPS and APS tungsten coatings. The domain of the model was a 2D image recorded using SEM. The image was thresholded based on the grey scale level in the micrographs. The dark phase represented the pores while the light phase represented tungsten. A finite element mesh was created over the image. It was observed that VPS coatings were well represented by the 2D model while APS coatings were not due to their complex microstructure [22]. A method to compensate for 3D pore structure effects was also proposed by choosing the maximum effective pore length for APS coatings and sectioning and shifting the model prior to performing the node temperature calculation [22].

A finite difference method-based approach was given by Dorvaux et al. [7] by developing software for the computation of the thermal conductivity of porous coatings from binary images of real material cross sections. This approach takes into account the complex morphology of the ceramic. It was used to determine the contribution of each pore family (globular pores, cracks etc.) to thermal insulating capabilities with the help of image analysis. It was concluded that major heat insulation is provided by the cracks oriented perpendicular to the heat flow direction [7]. This approach was used later as an alternative to diffusivity measurements for ranking coatings according to their heat insulation capacity with regard to the morphology [23]. It was also used to study the relationship between sintering effects and thermal conductivity increase and to study the effect of pressure on thermal conductivity of TBCs [24]. This approach was associated with a morphology generator to develop the software Tbctool (ONERA, France) for building a predictive tool which can be used to generate plasma-sprayed TBC-like morphology. This tool has been discussed in detail in Sect. 5.4.

Bartsch et al. [25] compared the results from finite difference (Tbctool) and finite element simulations for determining the thermal response from binary images of EB-PVD TBC microstructures. It was observed that the results are more accurate with decreasing mesh size and when the calculated section area of the microstructure image is larger. The finite element codes resulted in higher conductivity values compared to finite difference codes, for the same sections and similar mesh width. It was concluded that finite difference codes take less computer memory for calculations for larger models [25].

Kulkarni et al. [26] used SANS to study the effect of material feedstock characteristics (powder morphology) on the properties of plasma-sprayed YSZ coatings. A 2D finite element model representing the coating microstructure was built from the volumetrically averaged information available from the SANS data. The voids were divided into three families—interlamellar pores (assumed to be hexagonal in shape), intrasplat cracks and globular or irregular pores. Aspect ratios were assumed for each void family and the volume fractions, mean pore dimensions and orientations were taken from anisotropic multiple SANS data. The predicted thermal conductivity was higher than the experimentally measured values due to lack of information on the splat boundaries and pore size distribution [26].

Tan et al. [27] used processed SEM images of coating microstructure to generate a 2D finite element mesh and predict the effective thermal conductivity using a commercial finite element code for YSZ, molybdenum and NiAl coatings. It was observed that the modelling procedure was not sensitive to slight changes in the threshold level of the images. For YSZ coatings, the predicted thermal conductivity values were higher than the values measured by LFA, but they still followed the same trends, for both before and after annealing the coatings [27].

Liu et al. [28] developed a 3D finite element model which included spherical pores and unmelted particles, and ellipsoidal cracks, each inside a unit cell.

The unmelted particles were assumed to be completely debonded from the substrate. The unit cell model was used to predict the thermal insulation behaviour of different features on YSZ plasma-sprayed coatings.

Qunbo et al. [29] used digital image processing to create a finite element mesh over a real microstructure image to predict the thermo-mechanical properties of TBCs and concluded that transverse cracks in the coating are most significant for thermal insulation.

5.2   Young's Modulus

5.2.1   Analytical Models

Li et al. [30] developed an idealised model for estimating Young's modulus of thermal sprayed ceramic coatings consisting of the stacking of few micrometre-thick lamellae using circular plate bending theory. Two components of elastic strain of the coating under tensile stress were considered in the model, one related to the localised elastic strain at the regions of contact area between lamellae and the other related to the elastic bending of the lamellae between bonded regions. It was shown that the bending component becomes significant only when the percentage bonding ratio between lamellae becomes less than 40 %.

Sevostianov et al. [31] developed a model for Young's modulus as a function of microstructural parameters similar to the one described in Sect. 5.1.1 for thermal conductivity and proposed an equation as given below:

$$E_{tot} = E_d \left[1 + \frac{8(4-v_0)(1-v_0^2)}{3(2-v_0)} \frac{\alpha_v}{1-P} \right]^{-1} \tag{5.6}$$

where v_0 is Poisson's ratio of the material; other parameters are defined in earlier equations.

Azarmi et al. [32] investigated APS deposited alloy 625 coatings and compared Young's modulus measured by uniaxial tension test with several analytical models available for predicting Young's modulus of porous materials and a finite element modelling technique OOF (discussed in Sect. 5.3.2). A significant difference was observed between the predicted values by analytical models and experimentally measured values of the elastic modulus.

5.2.2   Numerical Models

Michlik et al. [33] used a FEM approach using a finite element package OOF pre-processor to generate the mesh and employ an in-house developed solver using extended finite element method (XFEM). The XFEM approach accounted for the existence of cracks in TBCs which enabled to study the effect of sintering of coating on elastic modulus.

Amsellem et al. [34, 35] used both 2D and 3D analyses of APS microstructures to determine Young's modulus of the coatings. Two-dimensional finite element meshes were created from SEM images of microstructures while 3D microstructures were obtained using X-ray microtomography (XMT). The limitations of both 2D and 3D modelling approaches were discussed and it was concluded that the 2D approach was suitable for characterising mechanical properties due to the low resolution of XMT [34]. Three-dimensional approach was assessed to be beneficial for studying the shape and orientation of pores as well as to simulate the stress concentration in the coating.

Apart from the numerical techniques discussed above, another finite element technique which has been used extensively to determine the thermal–mechanical properties of APS coatings is OOF, as described in the following section.

5.3 Finite Element Modelling

The over-simplification of porosity description in the analytical models limits the predictive capacity and the certainty for the complex interconnected porous structures. Numerical schemes seem to be more promising in such cases as they involve a material structure which is close to the real structure [7, 32]. In cases where the spatial geometry is too complex for the usage of analytical models, numerical modelling such as FEM and finite difference method (FDM) is beneficial.

Two important issues must be considered for numerical modelling of complex microstructural features present in APS coatings. First, the model sample volume must be large enough to contain sufficient microstructural details and hence represent the average property of the sample. Second, fine details of the microstructural features have to be captured within the model volume so that the actual properties are estimated precisely.

5.3.1 Basics of FEM and FDM

Both FEM and FDM are numerical approximation techniques to solve a set of equations which are typically not easily solvable analytically, for example due to complicated irregular geometry, complex material composition, etc. In both methods, the geometry is fractioned in small basic units, the problem is set up individually in each basic unit and then solved together to calculate the result. Only lines/squares/cubes can be used as basic units in FDM, whereas arbitrarily shaped basic units can be used in FEM thus enabling better and less complex description of the geometry. The FEM hence requires less computational power and has a better accuracy than FDM.

In addition, FEM is more suitable as an adaptive method since it easily facilitates for local refinements of the solution.

FEM is applied to most engineering problems either by writing a custom-made finite element program using softwares such as Matlab (MathWorks Inc., MA, USA) or by using an already developed commercial finite element program. Commercial finite element tools have a variety of built-in features for modelling different engineering problems but the cost associated with purchasing these tools and learning the way how to use them is usually high. Custom-made finite element programs might have a small cost initially but they handle a small variety of problems at a time and are less flexible. The solution of a finite element problem involves the construction of a matrix equation of type $AX = B$ where A is the stiffness matrix, X is the vector containing all the unknowns and B is the vector that depends on the boundary condition.

5.3.2   Image Based Finite Element Model

The possibility to determine a material's macroscopic properties based on its microstructure is of great importance to materials science. The object-oriented finite element analysis (OOF) code, developed at the Center for Computational and Theoretical Materials Science (CTCMS) at US National Institute of Standards and Technology (NIST), enables complex two-dimensional microstructures to be modelled using images of the actual microstructures. Its current version, OOF2, is available as public domain software [36]. This technique has been employed extensively in recent years to simulate thermo-mechanical properties of APS coatings.

Wang et al. [37] compared the models based on SANS and OOF for predicting thermal conductivity of plasma-sprayed YSZ coatings. The effect of splat boundaries on thermal conductivity was also discussed, and it was concluded that splat interfaces account for 25–60 % of the total reduction in conductivity, with larger values for higher splat interface density coatings [37].

Jadhav et al. [38] used OOF and an analytical model to predict thermal conductivity of APS and SPPS coatings and compared them with the experimental values measured by LFA. It was concluded that OOF model fitted better with experimental values compared to the analytical model and it captured the effect of real microstructures on the thermal conductivities more accurately [38]. The model was also used later to determine residual stresses in SPPS TBCs [39]. The low value of predicted residual stresses depicted that the TBC microstructures produced by SPPS were strain tolerant.

Bolot et al. [40] predicted thermal conductivity of AlSi/polyester coatings with two different numerical codes—TS2C (based on finite differences) and OOF. Some discrepancies between the values obtained were observed, why the code was developed further by changing the discretisation method which gave better results [41].

5.3.2.1 Modelling Procedure

OOF is very efficient in capturing the actual microscopic images and generating the required finite element meshes based on the colours present in the image [42, 43]. Though it is limited to 2D images, it can be very useful in determining macroscopic properties of materials consisting of more than one phase.

The model geometry information for analysis is obtained from the microstructure image which is used as an input. Different phases/features present in the microstructure image are then grouped based on their colour and/or contrast. Each group of pixels can be assigned a specific material property. The microstructure images are usually represented in greyscale. Even though OOF2 offers several tools for creating pixel groups of a greyscale image, it is simpler to modify the image before using it in OOF2. For example, if there are two phases present in the image, the image can be converted into binary format by thresholding to reduce the number of hues, thus making it more suitable for an automatic and repetitive method as shown in Fig. 5.1.

Rough skeletons are first generated on the image using an adaptive meshing routine. The skeleton is a grid that adapts to all pixel group boundaries present in the image. Initially individual elements may contain pixels corresponding to different individual phases. During the meshing procedure, the elements are refined and the nodes are moved so that the material interfaces are well defined. Based on the boundaries of the defined pixel groups present in the image, the skeleton is refined to capture the irregular boundaries marking the different phases of the microstructure. Finer elements are automatically generated near the interfaces to account for higher gradients.

There are several skeleton modifying options in OOF2, many dependent on the two types of 'energies' of the elements: shape energy and homogeneity energy [44]. The total energy (E) of an element is calculated with the following equation:

$$E = \alpha_h . E\_\mathrm{hom} + \left(1 - \alpha_h\right) . E\_\mathrm{shape} \tag{5.7}$$

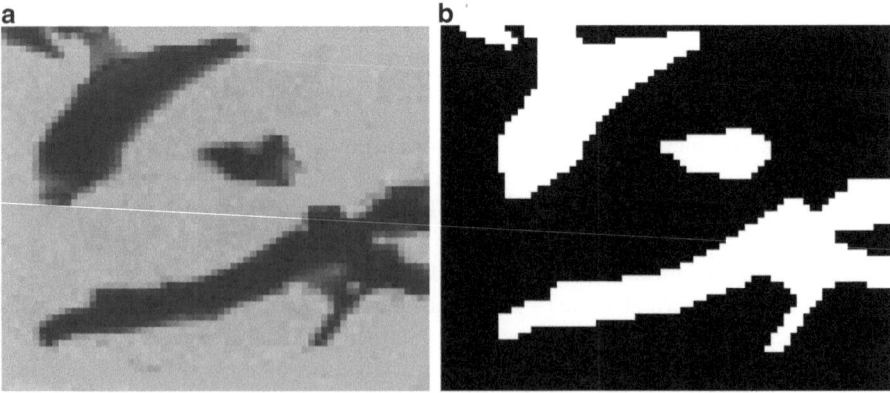

Fig. 5.1 A small portion of a microstructure image in (**a**) the original greyscale and (**b**) binary format

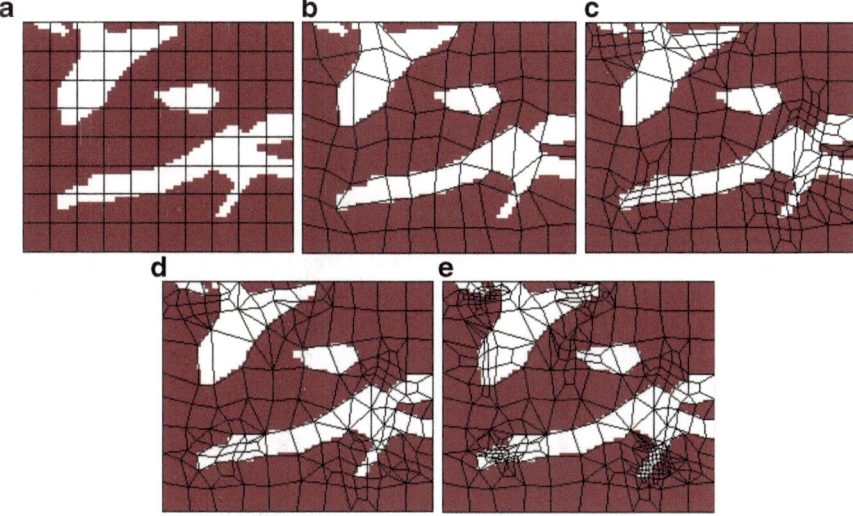

Fig. 5.2 A skeleton modification process in the FE software OOF based on the image in Fig. 5.1b

where *E_hom* is the definition of the homogeneity energy, *E_shape* is the definition of the shape energy and α_h is an adjustable parameter between 0 and 1. A high value of α_h is used when the priority is the homogeneity of elements, i.e. it is more important that the element nodes match the image boundaries rather than keeping a good shape.

An example of a skeleton modification process is shown in Fig. 5.2. Figure 5.2a shows an initial skeleton, without any consideration to the image boundaries. This skeleton is generated in the first step based on the maximum element size defined by the user. Figure 5.2b shows the skeleton after a number of *Anneal* iterations have been performed. The *Anneal* routine moves nodes randomly and accepts moves according to a given criterion [32]. In Fig. 5.2c the *Refine* method has been used, which splits elements into smaller pieces to make the skeleton correspond better to the boundaries [32]. In Fig. 5.2d, e several additional skeleton modifying steps, e.g., *Rationalise*, have been used to further optimise the skeleton. The *Rationalise* option fixes badly shaped elements by modifying them and adjacent elements or by removing them completely [43].

During the skeleton creation process, OOF2 displays a homogeneity index showing the average homogeneity level of all elements. This value should increase during the process, as the skeleton gets finer and more adapted to the boundaries. A high value of homogeneity index is desired which would imply that the different pixel groups are well defined by the created mesh. If the index reaches 1, all elements are 100 % homogeneous.

When the meshing procedure is completed, the elements are assigned the property of the material with the dominant pixel group present in each element to generate the final mesh. Once the mesh is created, the equation system and boundary conditions are defined over the model.

The main quality concern in OOF is capturing the required microstructural details present in the images. Since the fine details of phases/features present in the image depend on the quality of image, SEM images are preferred over optical microscope images. Lower resolution images also result in higher errors due to inefficient discretisation of the image during the thresholding procedure. Dorvaux et al. and Tan et al. have discussed the uncertainties arising due to the effects of image threshold level, image location across the sample cross section and image magnification and size on the final results [7, 27]. These factors must be kept in mind when using OOF for predicting macroscopic material properties. In general, an image covering a large area of cross section with high resolution should be used. Several images acquired across the sample cross section must be used so as to reduce the errors induced due to the effect of local features on predicted properties.

The quality of mesh generated over the group of pixels representing the different phases present in the microstructure image also has a major influence on the modelling results. It is found that minimising the maximum scale of element size is crucial in addition to setting a small minimum element size. Figure 5.3 shows the finite element mesh generated with coarse and fine element sizes. The red and white areas represent the two phases (pixel groups) present in the image. As shown in Fig. 5.3, the mesh with fine element is capable of capturing the small microstructural details which are missed out in the coarse mesh, since the material property for an element is assigned on the basis of the dominant pixel group present in that element. Thus, a finer mesh is required for higher accuracy of the predicted results.

An advantage of using OOF2 is that the meshes created in OOF2 can be saved and exported in different file formats, which enables the use of OOF2 meshes in other finite element software applications, such as ANSYS Workbench (ANSYS Inc., USA).

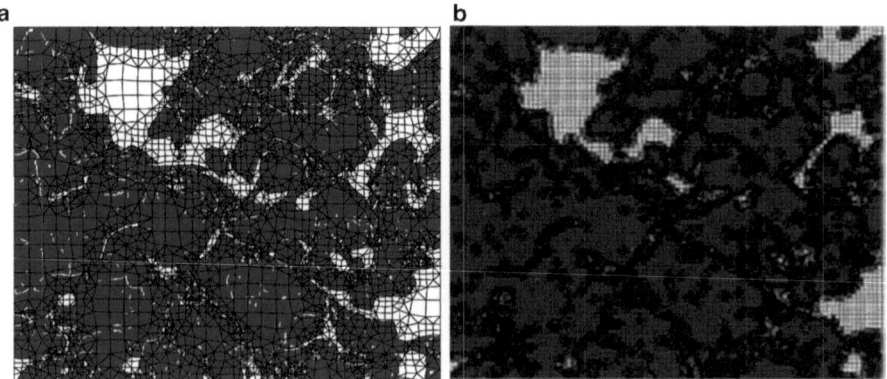

Fig. 5.3 Finite element mesh with different maximum and minimum element sizes. It can be clearly noticed that the mesh in (**b**) is able to capture much finer details compared to (**a**)

This is highly beneficial when more tedious calculations are necessary and/or additional simulation options are required. Another advantage is that once the meshing procedure is optimised for one of the microstructure images for a sample, the entire procedure can be automated. This can be done by saving the input instructions in a script form and changing only the text referring to input image file for other images for the sample. Thus, errors induced due to the operator can be reduced.

5.3.2.2   Property Evaluation

Once a finite element mesh has been created over the microstructure image, a simple set-up could be made to evaluate properties such as thermal conductivity and Young's modulus using the mesh as shown in Fig. 5.4.

Thermal conductivity (λ) can be evaluated by setting up a temperature difference (ΔT) across the top and bottom boundaries while keeping the other boundaries insulated to calculate the heat flux (dQ_y/dt) and then using the steady-state heat equation in 2D domain given as

$$dQ_y / dt = \lambda \cdot A_y \left(\Delta T / L_y \right) \tag{5.8}$$

where A_y is the cross-sectional area perpendicular to the y-axis and L_y is the thickness of the model domain along the y-axis as shown in Fig. 5.4.

Similarly Young's modulus (E) can be evaluated by setting up a prescribed deformation (δ_x) along the x-direction to calculate the average stress (σ_x) across the model and then using the stress–strain equation given as

$$\sigma_x = E \left(\delta_x / L_x \right) \tag{5.9}$$

where L_x is the thickness of the model domain along the x-axis as shown in Fig. 5.4.

Since the area representing the image is only a small fraction of the total cross-sectional area of the coating and the microstructural features might vary significantly from one part of the coating to the other, a number of images need to be analysed to achieve a statistical significance.

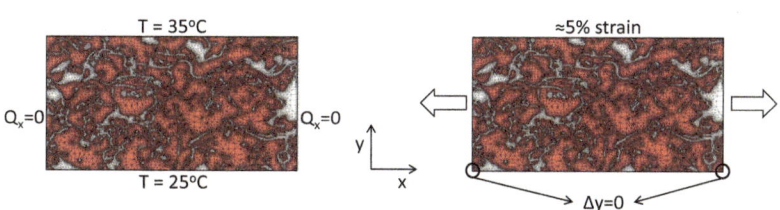

Fig. 5.4 Simple set-up to evaluate thermal conductivity (*left*) and Young's modulus (*right*)

5.4   Artificial Coating Morphology Generator

Tbctool software was developed by ONERA, France, within the Brite Euram HITS project (Project BE96-3226 HITS, task 1.4). It is an interactive computer program which was created especially for TBC development. As mentioned earlier, it has been used in the past to predict thermal conductivity of TBCs based on FDM. However, the major advantage of Tbctool is that it can be used as a predictive tool for generating plasma-sprayed TBC-like morphology. The software generates randomised microstructure images based on several pre-defined criteria. Figure 5.5 shows an artificially generated microstructure image using Tbctool simulating a plasma-sprayed coating morphology. The generation process is described in detail in the reference manual for Tbctool [45] and is described only briefly here.

The generation process is divided into four major sequences which are executed step by step: spheroids or globular pores generation, linked cracks or cracks connected to globular pores, free cracks and dust or fine scale pores. Each sequence has several parameters which can be defined using a statistical distribution function. The function can be defined using analytical distribution laws such as *Uniform*, *Normal*, *Log Normal* and *Custom* or tabulated laws with numerical definition. This statistical or probabilistic definition of parameters creates the randomness in the generation process. However, it must be kept in mind that these sequences can end up in an infinite loop, if too strict constraints are imposed on the generation parameters.

In the spheroids generation process, first the amount of spheroidal porosity needs to be defined. After that, the software chooses a set of particles from the 'particles library' on the basis of other specifications, such as particle area, elongation, orientation and compactness, so that the total area of the particles chosen becomes larger than

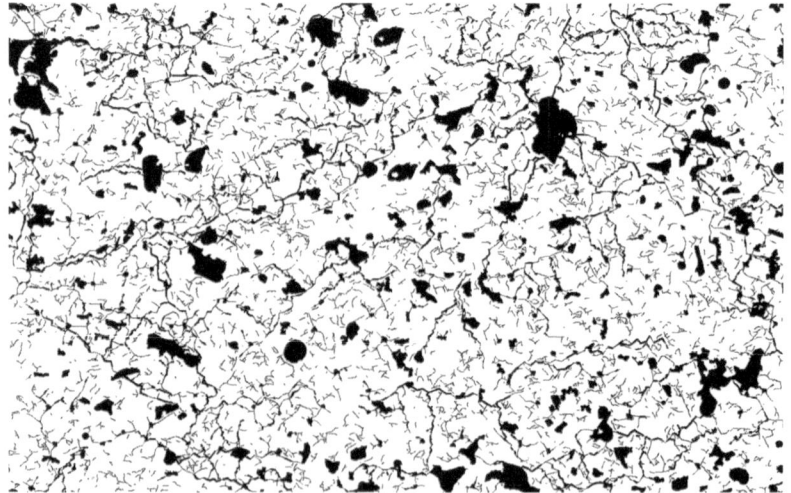

Fig. 5.5 An artificial image created using Tbctool morphology generator simulating a plasma-sprayed coating morphology

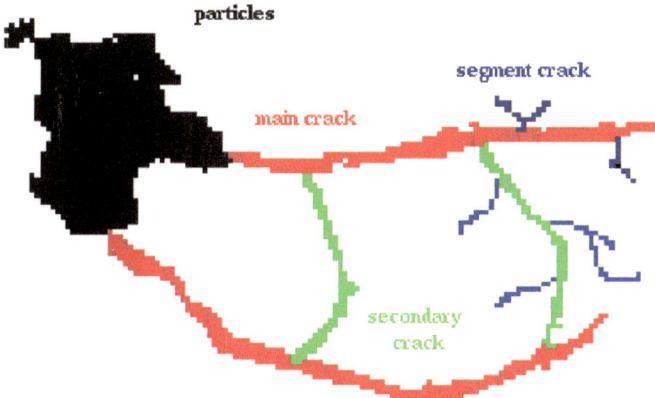

Fig. 5.6 Different crack families in Tbctool [45]

the amount of defined spheroidal porosity level. The particle library, which consists of thousands of particles or globular pores, is incorporated within the software. This particle library was developed by extracting features from the actual microscopic images of several APS coatings. Then the software assigns random positions to the set of chosen particles in the field on the basis of other specifications such as *overlapping rejection probability* and minimum distance between particles.

The linked cracks sequence is divided into three sub-sequences: main or primary cracks, secondary cracks and segment cracks. Figure 5.6 shows the different linked crack families. The primary cracks start from the spheroids and the secondary cracks start from the primary cracks, whereas the segment cracks may start from spheroids and/or other cracks. First, the amount of linked cracks porosity and the fractions of main linked cracks and secondary linked cracks need to be defined. All the cracks start from 'seeds', which are generated according to various specifications such as *seed birth probability* and *seed fraction*. All cracks are built according to several parameters related to crack shape (e.g. shape pulsation and shape damping) and crack size (e.g. crack length and crack thickness).

The free cracks generation sequence is very similar to the segment cracks generation process described above, the difference being in the nature of the seeds used. The dust generation sequence is the simplest one with the amount of dust porosity specified being generated on the basis of *dust distance* specified so that there is no overlapping of particles.

The main limitation of Tbctool morphology generator is that it uses geometric methods which are not directly related to the process physics. This means that the morphology generation parameters are somewhat arbitrarily related to the actual process parameters. The generation procedure widely uses random sub-processes which are difficult to optimise all together and might not have a physical significance. This limitation can also be used as an advantage since new coating morphologies can be fabricated virtually and explored for further analysis. However, in this case, the input parameters need to be linked to process parameters to ensure that the generated morphology is possible to achieve experimentally as well. Another limitation is the finite size of the particle library, so all types of particles cannot be generated using Tbctool, especially if one considers new coating materials. Comparisons between real and artificially generated microstructures and a methodology to design an optimised microstructure using Tbctool are presented in the next section.

5.5   Recent Work

In the first part of this work, the aim was to establish the fundamental understanding of relationships between coating microstructure and thermal conductivity of TBCs. A range of coating architectures was investigated including high purity YSZ, dysprosia-stabilised zirconia and dysprosia-stabilised zirconia with porosity former. The microstructures were examined both on as-sprayed samples and on heat treated samples. First, statistical modelling was used as a 'screening' method to determine the important microstructure parameters influencing the thermal conductivity. The most interesting finding in this analysis was that pores and cracks in contact seem to have significant influence on thermal conductivity compared to free pores and free cracks. As this model does not consider the actual physical phenomenon, the results from this analysis were validated by FEM developed with OOF using microstructure images obtained by SEM which provided physical verification of the statistical results. The microstructure exhibiting the best performance in this study is shown in Fig. 5.7 where the pores and cracks in contact are indicated. The results, although

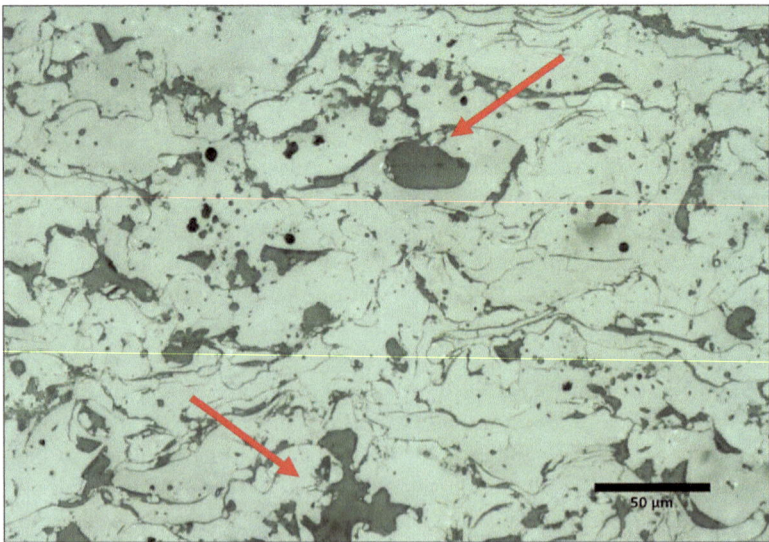

Fig. 5.7 Microstructure cross-section image indicating pores and cracks in contact [1]

tentative in nature, indicate that the combination of the two modelling approaches seems to be a feasible approach to understand relationships between coating microstructure and thermal conductivity of TBCs. The effect of pores and cracks in contact on TBC properties was investigated further in this work.

To design new morphologies, Tbctool was first verified on real coatings. First, Tbctool was used to generate artificial microstructure images using the input parameters determined by image analysis of real microstructure images obtained using SEM. These artificially generated images were used as an input to the finite element model developed using OOF to predict thermal conductivity. The artificially generated images were verified by two methods. First, by comparing their image analysis data with SEM images, and second, by comparing the predicted thermal conductivity values with the predicted thermal conductivity values from SEM images and the thermal conductivity values obtained from laser flash experiments. It was observed that the finite element model ranked the coating systems considered the same way in terms of thermal conductivity as the experimental values. Also, same ranking was observed for the artificial images generated. These tentative results indicate that the images generated by the coating morphology generator Tbctool were similar to the real coatings. Figure 5.8 shows microstructure cross-section images of one such TBC topcoat analysed in this work taken by SEM and artificially generated by Tbctool. It must be noted that these images might not be lookalike but they are similar from a statistical point of view based on the results obtained above. These results indicate that the combination of OOF and Tbctool can be used as a powerful approach to design coatings. Hence, this approach was used in further work as described below to develop a low thermal conductivity and long lifetime TBC.

Once the finite element model developed using OOF and the artificial coating morphology generator Tbctool were verified, a combined empirical and numerical approach was utilised to develop a novel thermal barrier coating. The intention was to design a coating system which could be implemented in industry within the next three years. Different morphologies of ceramic topcoat were evaluated: using dual layer systems and polymers to generate porosity. Dysprosia-stabilised zirconia was also included in this study as a topcoat material along with the state-of-the-art YSZ using

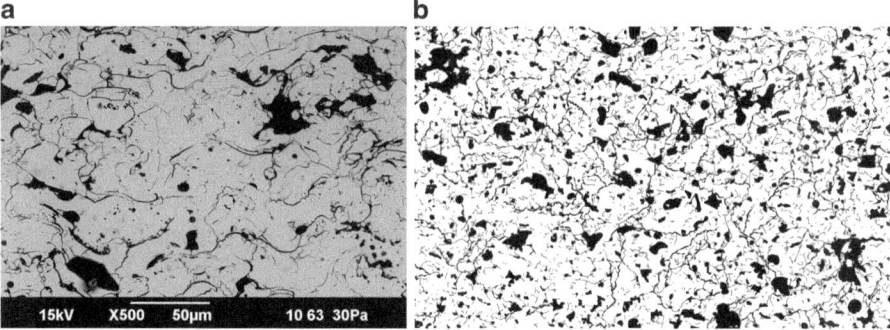

Fig. 5.8 Microstructure cross-section images of a TBC topcoat (**a**) taken by SEM and (**b**) artificially generated by Tbctool [2]

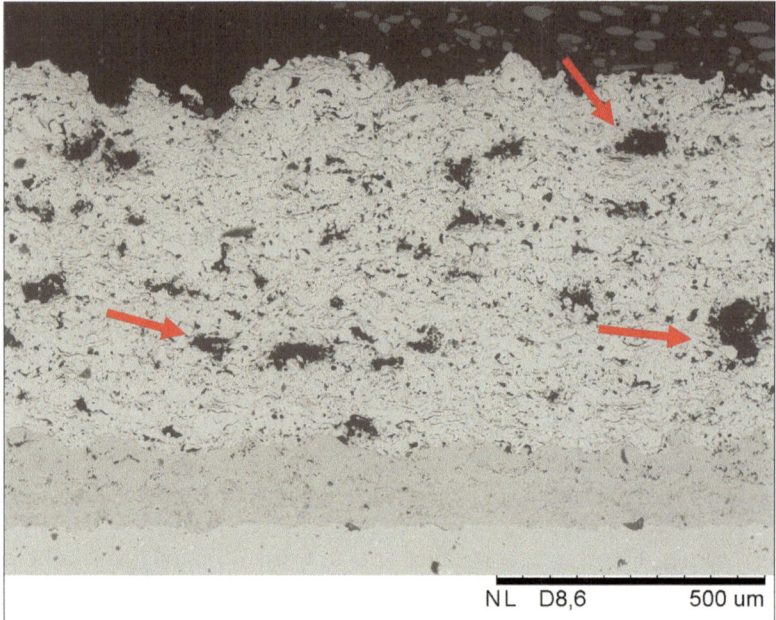

Fig. 5.9 Microstructure image of the porosity former coating indicating the large globular pores with connected cracks [3]

high purity powders. In the experiments it was shown that a dysprosia-doped high purity zirconia coating with large globular pores with connected cracks created due to the inclusion of polymers exhibited the best performance, i.e. lowest thermal conductivity and longest fatigue lifetime. The coating microstructure is shown in Fig. 5.9 where the large globular pores with connected cracks are indicated. The finite element model was shown to be capable of catching the trend in different coating lifetime through predicted Young's modulus values thus making it possible to optimise a coating design using the model.

To achieve further understanding of why this coating exhibited best performance, Tbctool was utilised to design three particular kinds of microstructure images containing only pores, only free cracks with pores and connected cracks with pores as shown in Fig. 5.10. The porosity level in all these three images was chosen to be the same, even though they comprised different microstructural features; the porosity level being similar to the images analysed earlier for porosity former coatings. It was observed that pores and connected cracks result in lowest Young's modulus and thermal conductivity, which also suggests an explanation for the high performance of porosity former coatings. It also shows that low Young's modulus and thermal conductivity cannot be achieved by only producing large pores, since that image resulted in a much higher values. The work implies that the combined empirical and numerical approach is an effective tool for developing low thermal conductivity coatings with enhanced lifetime. It was shown that a low conductivity coating with twice the lifetime compared to the industrial reference today could be produced.

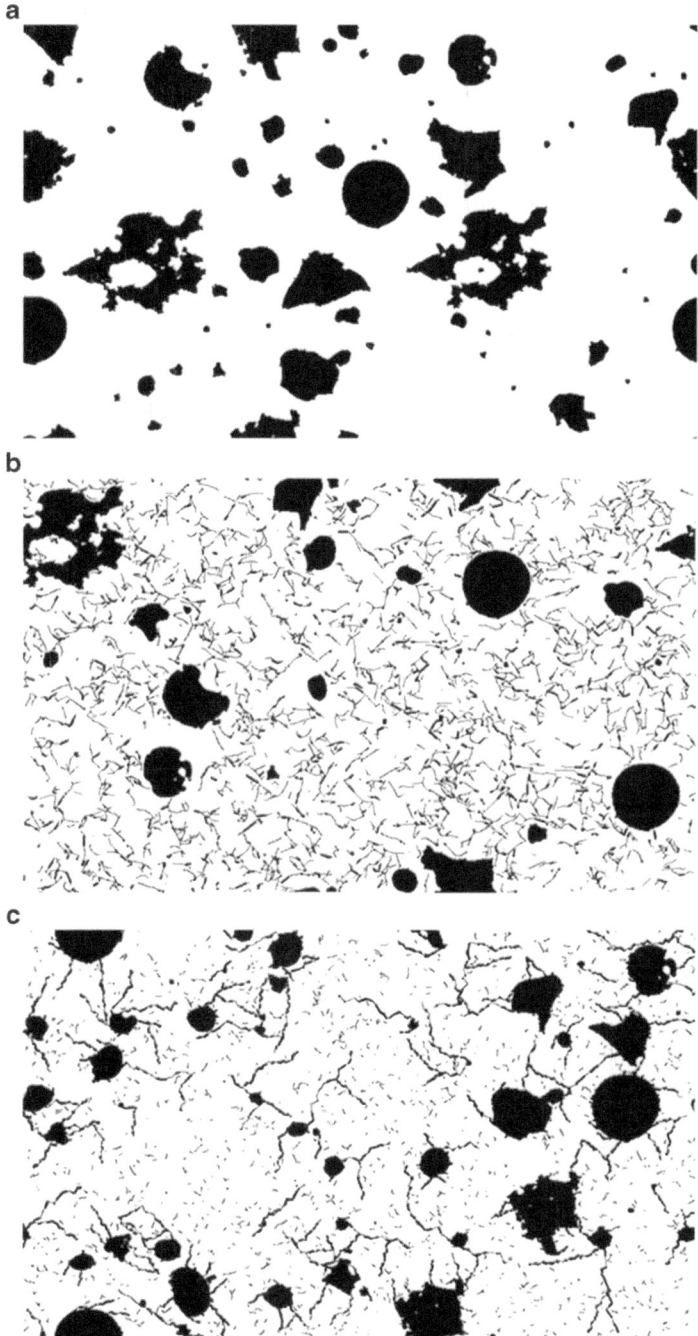

Fig. 5.10 Microstructure images generated using Tbctool representing (**a**) only pores, (**b**) free cracks with pores and (**c**) connected cracks with pores [3]

To further verify these results as well as the modelling approach, a design of
experiments (DoE) was conducted by varying selected spray parameters, and the finite
element model was used to predict thermal conductivity and Young's modulus. The
aim was to investigate fundamental relationships between microstructure and thermal–
mechanical properties and obtain an optimised coating microstructure. The lowest
thermal conductivity and Young's modulus were shown by a coating exhibiting large
globular pores with connected cracks in the microstructure as shown in Fig. 5.11a.

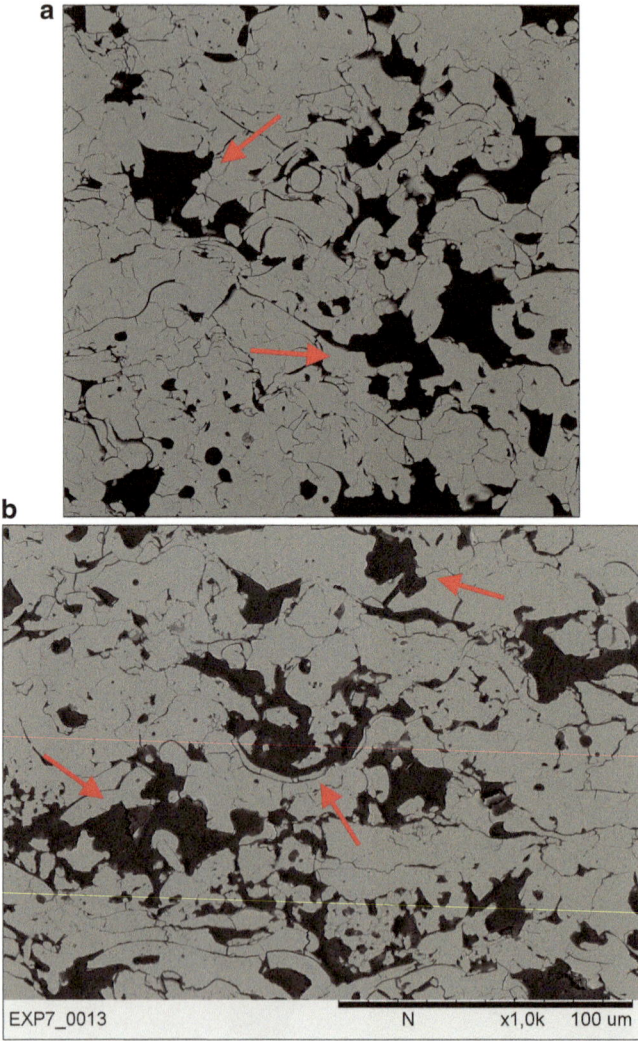

Fig. 5.11 Microstructure cross-section image indicating large pores with connected cracks
(**a**) coating showing best performance based on the design of experiments, and (**b**) optimised coating
sprayed with another spray gun [4]

For the purpose of verifying the modelling results, this microstructure was sprayed along with a reference coating with a new spray gun. Thermo-mechanical properties were predicted using the same finite element model while thermal conductivity and thermo-cyclic fatigue lifetime were measured experimentally. The coating microstructure with large globular pores with connected cracks shown in Fig. 5.11b performed better compared to the reference coating sprayed with standard settings before optimisation. In this way, the modelling results obtained in the first part of this work were verified. This work shows that the modelling approach in combination with experiments can be used as a powerful tool to achieve optimised microstructures. This approach is both time saving and cost effective.

References

1. Tano I, Gupta M, Curry N, Nylén P, Wigren J (2010) Relationships between coating microstructure and thermal conductivity in thermal barrier coatings: a modelling approach. Proceedings of the International Thermal Spray Conference, May 3–5, 2010 (Singapore), DVS Media, pp 66–70
2. Gupta M, Nylén P (2011) Design of low thermal conductivity thermal barrier coatings by finite element modelling. In: Sudarshan TS, Beyer E, Berger L-M (eds) Surface Modifications Technologies XXIV, Sep 7–9, 2010 (Dresden). pp 353–365
3. Gupta M, Nylén P, Wigren J (2013) A modelling approach to microstructure-property relationships in thermal barrier coatings. J Ceram Sci Technol 4(2):85–92
4. Gupta M, Curry N, Markocsan N, Nylén P, Vaßen R (2013) Design of next generation thermal barrier coatings- experiments and modelling. Surf Coat Technol 220:20–26
5. Pawlowski L, Fauchais P (1992) Thermal transport properties of thermally sprayed coatings. Int Mater Rev 37(6):271–289
6. Golosnoy IO, Tripas SA, Clyne TW (2005) An analytical model for simulation of heat flow in plasma-sprayed thermal barrier coatings. J Therm Spray Technol 14(2):205–214
7. Dorvaux JM, Lavigne O, Mevrel R, Poulain M, Renollet Y, Rio C (1998) Modelling the thermal conductivity of thermal barrier coatings. Proceedings of the 85th Meeting of the AGARD Structures and Materials Panel, Oct 15–16, 1997 (Aalborg, Denmark), NATO-AGARD-R-823, pp 13-1–13-10
8. Wagh AS (1993) Porosity dependence of thermal conductivity of ceramics and sedimentary rocks. J Mater Sci 28(14):3715–3721
9. Cernuschi F, Ahmaniemi S, Vuoristo P, Mäntylä T (2004) Modelling of thermal conductivity of porous materials: application to thick thermal barrier coatings. J Eur Ceram Soc 24(9): 2657–2667
10. Schlichting KW, Padture NP, Klemens PG (2001) Thermal conductivity of dense and porous yttria-stabilized zirconia. J Mater Sci 36(12):3003–3010
11. Klemens PG (1991) Thermal conductivity of inhomogeneous media. High Temp High Press 23(3):241–248
12. Hasselman DPH (1978) Effect of cracks on thermal conductivity. J Compos Mater 12: 403–407
13. McPherson R (1984) A model for the thermal conductivity of plasma-sprayed ceramic coatings. Thin Solid Films 112(1):89–95
14. Li CJ, Ohmori A (2002) Relationships between the microstructure and properties of thermally sprayed deposits. J Therm Spray Technol 11(3):365–374
15. Boire-Lavigne S, Moreau C, Saint-Jacques RG (1995) The relationship between the microstructure and thermal diffusivity of plasma-sprayed tungsten coatings. J Therm Spray Technol 4(3):261–267

16. Bjorneklett A, Haukeland L, Wigren J, Kristiansen H (1994) Effective medium theory and the thermal conductivity of plasma-sprayed ceramic coatings. J Mater Sci 29(15):4043–4050
17. Sevostianov I, Kachanov M (2000) Anisotropic thermal conductivities of plasma-sprayed thermal barrier coatings in relation to the microstructure. J Therm Spray Technol 9(4):478–482
18. Sevostianov I, Kachanov M, Ruud J, Lorraine P, Dubois M (2004) Quantitative characterization of microstructures of plasma-sprayed coatings and their conductive and elastic properties. Mater Sci Eng A 386(1–2):164–174
19. Sevostianov I, Kachanov M (2009) Elastic and conductive properties of plasma-sprayed ceramic coatings in relation to their microstructure: an overview. J Therm Spray Technol 18(5–6):822–834
20. Lu TJ, Levi CG, Wadley HNG, Evans AG (2001) Distributed porosity as a control parameter for oxide thermal barriers made by physical vapor deposition. J Am Ceram Soc 84(12): 2937–2946
21. Cipitria A, Golosnoy IO, Clyne TW (2009) A sintering model for plasma-sprayed zirconia TBCs. Part I: free-standing coatings. Acta Mater 57(4):980–992
22. Hollis KJ (1995) Pore phase mapping and finite-element modeling of plasma sprayed tungsten coatings. In: Berndt CC, Sampath S (ed) Advances in thermal spray science & technology. Sept 11–15, 1995 (Houston, TX), ASM International, pp 403–409
23. Saint-Ramond B (2001) HITS: high insulation thermal barrier coating systems. Air Space Europe 3(3–4):174–177
24. Poulain M, Dorvaux JM, Lavigne O, Mévrel R, Renollet Y, Rio C (2002) Computation of thermal conductivity of porous materials applications to plasma sprayed TBCs. In: Advanced thermal barrier coatings and titanium aluminides for gas turbines, June 17–19, 2002 (Bonn, Germany), Turbomat
25. Bartsch M, Schulz U, Dorvaux JM, Lavigne O, Fuller ER Jr, Langer SA (2003) Simulating thermal response of EB-PVD thermal barrier coating microstructures. Ceram Eng Sci Proc 24(3):549–554
26. Kulkarni A, Wang Z, Nakamura T, Sampath S, Goland A, Herman H, Allen J, Ilavsky J, Long G, Frahm J, Steinbrech RW (2003) Comprehensive microstructural characterization and predictive property modeling of plasma-sprayed zirconia coatings. Acta Mater 51(9):2457–2475
27. Tan Y, Longtin JP, Sampath S (2006) Modeling thermal conductivity of thermal spray coatings: comparing predictions to experiments. J Therm Spray Technol 15(4):545–552
28. Liu FR, Zeng KL, Wang H, Zhao XD, Ren XJ, Yu YG (2010) Numerical investigation on the heat insulation behavior of thermal spray coating by unit cell model (#1182). Proceedings of the International Thermal Spray Conference, May 3–5, 2010 (Singapore), DVS Media, 2010, pp 794–798
29. Qunbo F, Fuchi W, Lu W, Zhuang M (2009) Microstructure-based prediction of properties for thermal barrier coatings, thermal spray 2009: expanding thermal spray performance to new markets and applications. Marple BR, Hyland MM, Lau Y-C, Li C-J, Lima RS, Montavon G (eds), May 4–7, 2009 (Las Vegas, Nevada), ASM International, pp 46–50
30. Li C, Ohmori A, McPherson R (1997) The relationship between microstructure and Young's modulus of thermally sprayed ceramic coatings. J Mater Sci 32:997–1004
31. Sevostianov I, Kachanov M (2001) Plasma-sprayed ceramic coatings: anisotropic elastic and conductive properties in relation to the microstructure; cross-property correlations. Mater Sci Eng A 297(1–2):235–243
32. Azarmi F, Coyle T, Mostaghimi J (2009) Young's modulus measurement and study of the relationship between mechanical properties and microstructure of air plasma sprayed alloy 625. Surf Coat Technol 203:1045–1054
33. Michlik P, Berndt C (2006) Image-based extended finite element modeling of thermal barrier coatings. Surf Coat Technol 201:2369–2380
34. Amsellem O, Madi K, Borit F, Jeulin D, Guipont V, Jeandin M, Boller E, Pauchet F (2008) Two-dimensional (2D) and three-dimensional (3D) analyses of plasma-sprayed alumina microstructures for finite-element simulation of Young's modulus. J Mater Sci 43:4091–4098

35. Amsellem O, Borit F, Jeulin D, Guipont V, Jeandin M, Boller E, Pauchet F (2012) Three-dimensional simulation of porosity in plasma-sprayed alumina using microtomography and electrochemical impedance spectrometry for finite element modeling of properties. J Therm Spray Technol 21(2):193–201
36. http://www.ctcms.nist.gov/oof/oof2/. Accessed 24 Feb 2015
37. Wang Z, Kulkarni A, Deshpande S, Nakamura T, Herman H (2003) Effects of pores and interfaces on effective properties of plasma sprayed zirconia coatings. Acta Mater 51(18):5319–5334
38. Jadhav AD, Padture NP, Jordan EH, Gell M, Miranzo P, Fuller ER Jr (2006) Low-thermal-conductivity plasma-sprayed thermal barrier coatings with engineered microstructures. Acta Mater 54(12):3343–3349
39. Jadhav AD, Padture NP (2008) Mechanical properties of solution-precursor plasma-sprayed thermal barrier coatings. Surf Coat Technol 202:4976–4979
40. Bolot R, Seichepine JL, Vucko F, Coddet C, Sporer D, Fiala P, Bartlett B (2008) Thermal conductivity of AlSi/Polyester abradable coatings. In: Lugscheider E (ed) Thermal spray crossing borders, June 2–4, 2008 (Maastricht, The Netherlands), ASM International. pp 1056–1061
41. Bolot R, Seichepine JL, Qiao JH, Coddet C (2011) Predicting the thermal conductivity of AlSi/polyester abradable coatings: effects of the numerical method. J Therm Spray Technol 20(1–2):39–47
42. Langer SA, Fuller ER Jr, Carter WC (2001) OOF: an image-based finite-element analysis of material microstructures. Comput Sci Eng 3(3):15–23
43. Reid ACE, Langer SA, Lua RC, Coffman VR, Haan S–I, García RE (2008) Image-based finite element mesh construction for material microstructures. Comput Mater Sci 43(4):989–999
44. Busso EP, Lin J, Sakurai S, Nakayama M (2001) A mechanistic study of oxidation-induced degradation in a plasma-sprayed thermal barrier coating system. Part I: model formulation. Acta Mater 49:1515–1528
45. Tbctool manual, Tbctool documentation

Chapter 6
Modelling of Interface Roughness in TBCs

6.1 Simplified Interface Roughness Modelling

The influence of topcoat–bondcoat interface roughness on thermo-mechanical stresses is of high relevance for understanding the failure mechanisms as well as assessing the lifetime of TBCs. Analysing the effect of interface roughness on stresses could be a tedious and extremely complicated task to perform experimentally. Therefore, numerical modelling techniques are commonly used to understand these fundamental relationships. In most of the earlier works, a 2D or 3D sinusoidal wave profile has been chosen as a simplification to represent the topcoat–bondcoat interface [1–16] to analyse the stress distribution in TBCs.

A crack propagation model was proposed by Bäker [2], Ranjbar-Far et al. [7] and Bialas [11] to analyse the stress distribution in TBCs by using a 2D wave profile to represent the topcoat–bondcoat interface. Ranjbar-Far et al. used uniform as well as non-uniform amplitudes to represent the wave profile and an inhomogeneous top-coat layer with an artificial lamellar structure. It was concluded that the cracking depends mainly on the interface morphology and the thickness of TGO layer.

Failure mechanisms were analysed by Evans [17] with the help of a 2D finite element model representing the interface as 2D sine wave profile. Lifetime prediction models were made by Vaßen et al. [18], Wei et al. [6], Brodin et al. [14] and He et al. [16] based on the growth of delamination cracks using a 2D sine wave profile for modelling the interface. The lifetime model by Vaßen et al. was based on the stress distribution analysis model developed by Ahrens et al. [3].

Kyaw et al. [5] used a 2D sine wave profile to represent the interface and incorporated the temperature dependency of material properties and sintering of topcoat in the finite element model. Qi et al. [8] used a 2D sine wave profile to represent the interface and attributed the stress distribution to the interface roughness and thermal mismatch. Asghari et al. [9] used a 2D semi-circular profile to represent the interface with a non-linear material model for topcoat implementing the effect of sintering. Hermosilla et al. [10] and Busso et al. [4] developed a couple microstructural–mechanical model

© The Author(s) 2015
M. Gupta, *Design of Thermal Barrier Coatings*, SpringerBriefs in Materials,
DOI 10.1007/978-3-319-17254-5_6

using a 2D wave profile to represent the interface. The TGO growth was determined by a diffusion model in these studies.

Bednarz [12] did an extensive work in this field by using a 2D profile with different shapes such as sinusoidal, semi-circular and elliptical. The major conclusion of this work was that the most significant parameters affecting the stress development during thermal cycling are time-dependent material properties of topcoat and TGO, TGO growth behaviour and interface roughness. Jinnestrand et al. [1] developed a model using a 3D sine wave profile to represent the interface to analyse the stress distribution.

Ghafouri-Azar et al. [19] and Klusemann et al. [20] made a 2D model to analyse the residual stresses in tungsten carbide–cobalt (WC-Co) coatings sprayed by HVOF method. Micrograph images were used to define the coating as well as substrate geometry for the finite element model including the coating–substrate interface.

As it could be observed from the above references, the general research trend in the modelling approach in this field has been to develop the material models for individual layers and simulation conditions so as to represent the real simulation conditions in a more precise way [5]. However, a simplified roughness profile might not lead to precise predictions as it does not incorporate the complex surface topography created during thermal spraying.

6.2   Real Interface Roughness Modelling

In a recent work by Nordhorn et al., finite element analyses of realistic 3D models captured using laser microscopy and multiple 2D models based on different interface approximation functions were performed to predict the thermally induced stress fields [21]. The objective was to simulate the 3D reference stress fields using 2D models so as to reduce the required computational time to assess the stress fields. It was concluded in this work that the 3D model was too complex with respect to the geometric interface features to be replaced by a single 2D approximating function. However, the 2D models could correctly reflect essential TGO growth-related features such as stress inversion.

In Sects. 6.2.1 and 6.2.2 below, the methodology undertaken by the author to model the interface roughness in 2D and 3D has been described in detail.

6.2.1   Two-Dimensional Approach

The 2D simulations were prepared using cross-sectional microstructure SEM images of isothermal heat treated samples with different time span in the furnace. The varying exposure time resulted in altered TGO growth of the samples. Three TGO thickness stages were considered in this work, as-sprayed, 100- and 200-h heat treatments so that a significant difference can be observed. Five images per specimen were analysed for this study for statistical significance.

The SEM images used in 2D modeling were captured carefully so as to include sections of the TGO, topcoat and bondcoat layers. These images were processed with image editing software Adobe Photoshop CS6 (Adobe Systems Incorporated, USA) as follows. Additional topcoat and bondcoat layers were added to the image to achieve desired thicknesses of individual layers as shown in Fig. 6.1. All porosities and cracks present in the SEM image were eliminated while the TGO interface profile was kept intact. Each layer was given an individual colour.

OOF2 was used to generate a mesh over the processed microstructure image and thereby enable a 2D simulation in the finite element software ANSYS. Before mesh generation, different pixel groups are created in OOF2 on the basis of colours present in the image, and material properties were assigned to these colour groups. A fine mesh is generated near the boundaries between the layers, and coarser elements are created where the homogeneity of elements was easily attained, as shown in Fig. 6.1. The mesh generated in OOF2 was then exported to ANSYS. The substrate layer was added subsequently in ANSYS Workbench as schematically shown in Fig. 6.1 to achieve the desired thicknesses of individual layers. A coarse mesh was generated in ANSYS over the substrate layer and merged with the imported mesh to generate the final mesh.

Once a finite element model is created using the procedure described above, it can be used for determining the stresses generated due to the interface profile by setting up simulation conditions.

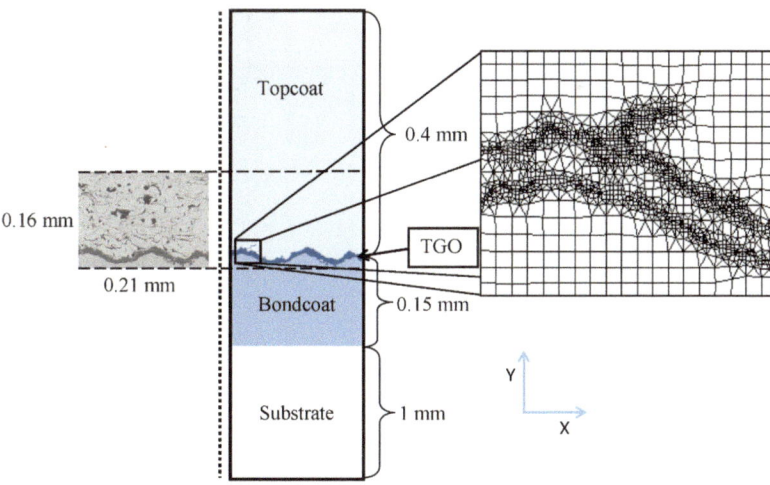

Fig. 6.1 SEM microstructure image of a TBC cross section (*leftmost*) is used to create the model shown in the centre. The substrate has been cropped. A part of the model illustrating the generated finite element mesh is shown to the *right* with varying element sizes in the TGO and adjacent boundaries [22]

6.2.2   Three-Dimensional Approach

The surface coordinate file obtained from 3D surface scanning techniques mentioned in Sect. 4.4 was used as an input for the 3D modelling approach. The bondcoat surfaces were scanned with white light interferometry technique (MicroXAM, ADE Phase Shift Technology, USA) before spraying the topcoat layer. The scanned data were stored in the form of a text file with a list of coordinates. The coordinate files were used to generate surfaces representing the curvature of the TGO and the adjacent faces of the bondcoat and the topcoat. Different TGO thicknesses were considered in the 3D simulations similar to the 2D simulations to analyse the effect of TGO growth on stresses.

The coordinate files representing the bondcoat topography obtained from white light interferometry were pre-processed in MATLAB R2010b (MathWorks, USA) to obtain an area of 50 μm × 50 μm to avoid tedious calculations as well as for compatibility with the computer-aided design (CAD) software, NX Unigraphics 7.5 (Siemens PLM Software, USA). The surface generated in the CAD software was used to represent both the bondcoat–TGO and the TGO–topcoat interfaces obtained by translating the surface in the direction perpendicular to the coating deposition direction. Other layers in the TBC system were then extruded to create appropriate thicknesses as shown in Fig. 6.2.

The model created in the CAD software was then imported to ANSYS where the finite element was generated. The mesh concentration was kept dense near the TGO and coarse at the areas with lower stresses, as shown in Fig. 6.2. The topcoat and the

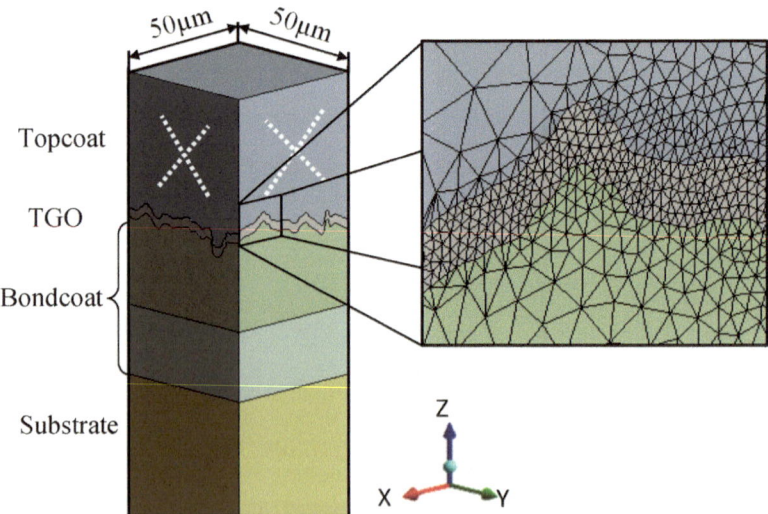

Fig. 6.2 Representation of the 3D model with cropped substrate and topcoat to the *left* and an enlargement of the TGO interface illustrating the varying mesh density present in the models to the *right* [22]

bondcoat were split into two bodies each to control the mesh density. Once a 3D finite element model is created using the procedure described, it can be used for determining the stresses generated due to the interface profile by setting up simulation conditions.

6.3 Results

By implementing the procedures described in Sects. 6.2.1 and 6.2.2, finite element modelling was used to study the residual stress profile in the topcoat–bondcoat interface using real surface topographies [22]. An attempt was made in this work to overcome the limitation of simplified representation of topcoat–bondcoat interface used conventionally by using real surface topographies so as to represent the interface in a more accurate way. The influence of topcoat–bondcoat interface on the induced stresses in topcoat which eventually affect lifetime was studied by using time to stress inversion as an indicator as per the stress inversion theory described in Sect. 3.4.2. The differences in functional performance between APS and HVOF bondcoat samples observed in an earlier work were evaluated with both 2D and 3D simulations [23]. Both 2D and 3D simulations were shown to verify the previously formulated stress inversion theory established using simplified sinusoidal curves. It was observed that the stress inversion from compressive to tensile stresses occurred earlier in the topcoat–bondcoat interface for the HVOF samples which could be a reason for an earlier failure of these samples in lifetime testing.

The bondcoat surface profiles captured using white light interferometry for the HVOF and APS samples are shown in Fig. 6.3. It was remarked based on the simulation results that possible unmelted particles present on the HVOF bondcoat surface could increase the overall stresses in the topcoat which could also contribute to earlier failure in lifetime testing. This could also be differentiated with the newly formulated ISO 25178 feature parameters shown in Table 6.1 derived from the segmentation motifs of these surface profiles.

The height parameter Sa which calculates the arithmetic mean height in 3D (corresponding to Ra in 2D) was 11.2 μm for HVOF and 11.1 μm for APS, which would suggest similar roughness level. However, from the feature parameters given in Table 6.1 based on segmentation motifs, it can be easily noticed that the surface topography is significantly different. The higher Spd value suggests that APS bondcoat has higher density of peaks compared to the HVOF bondcoat, and the higher Spc value suggests that the arithmetic mean curvature of these peaks is also higher. This implies that the APS bondcoat has high density of sharp and steep peaks while HVOF bondcoat has relatively flat and large surface area hills, which is also confirmed by the high Sha value for HVOF bondcoat. This conclusion from the quantitative measurements can also be easily visualised in Fig. 6.3.

These results support the argument that better characterisation of the surface rather than the traditional Ra value is required for a fundamental understanding. It was concluded that the modelling approach using real topographies can be a tool to

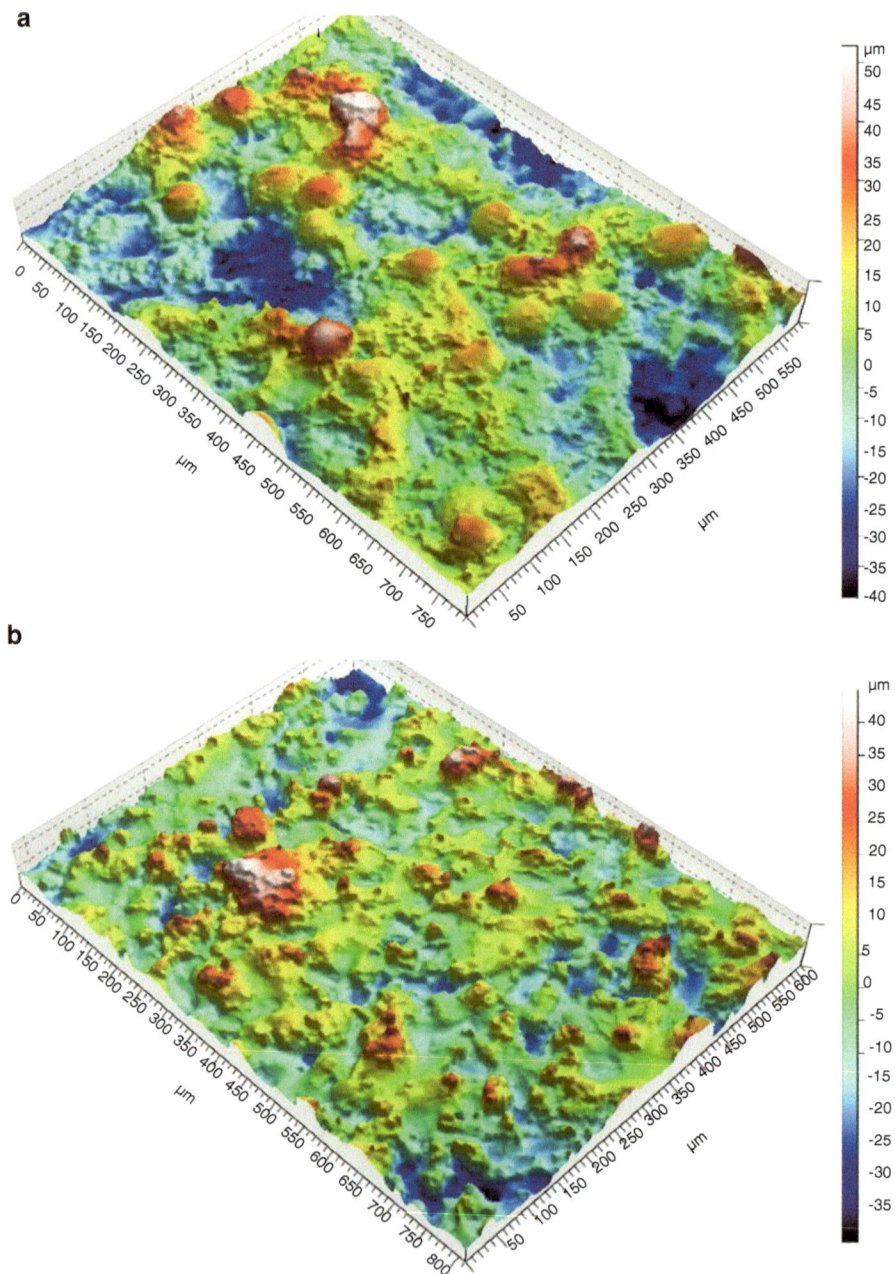

Fig. 6.3 Bondcoat surface profiles captured using white light interferometry technique for (**a**) HVOF and (**b**) APS samples [22]

Table 6.1 Some of the ISO 25178 feature parameters calculated for HVOF and APS bondcoat samples [22]

Parameter (units)	Parameter description	HVOF	APS
Spd $(1/mm^2)$	Density of peaks (number of peaks per unit area)	49.5	135
Spc $(1/mm)$	Arithmetic mean peak curvature (arithmetic mean of the principal curvatures of peaks within a definition area)	1,140	2,844
Sha (mm^2)	Mean hill area (average area of the hills connected to the edge at a particular height)	0.0194	0.00699

attain valuable insight into the stress distribution in topcoat–bondcoat interfaces. The results indicated that optimisation of bondcoat roughness can be realised through this technique.

References

1. Jinnestrand M, Sjöström S (2001) Investigation by 3D FE simulations of delamination crack initiation in TBC caused by alumina growth. Surf Coat Technol 135:188–195
2. Bäker M (2012) Finite element simulation of interface cracks in thermal barrier coatings. Comput Mater Sci 64:79–93
3. Ahrens M, Vaßen R, Stöver D (2002) Stress distributions in plasma-sprayed thermal barrier coatings as a function of interface roughness and oxide scale thickness. Surf Coat Technol 161:26–35
4. Busso EP, Lin J, Sakurai S, Nakayama M (2001) A mechanistic study of oxidation-induced degradation in a plasma-sprayed thermal barrier coating system. Part I: model formulation. Acta Mater 49:1515–1528
5. Kyaw S, Jones A, Hyde T (2013) Predicting failure within TBC system: Finite element simulation of stress within TBC system as affected by sintering of APS TBC, geometry of substrate and creep of TGO. Eng Fail Anal 27:150–164
6. Wei S, Fu-chi W, Qun-bo F, Zhuang M (2013) Lifetime prediction of plasma-sprayed thermal barrier coating systems. Surf Coat Technol 217:39–45
7. Ranjbar-Far M, Absi J, Mariaux G (2012) Finite element modeling of the different failure mechanisms of a plasma sprayed thermal barrier coatings system. J Therm Spray Technol 21(6):1234–1244
8. Qi H-Y, Yang X-G (2012) Computational analysis for understanding the failure mechanism of APS–TBC. Comput Mater Sci 57:38–42
9. Asghari S, Salimi M (2010) Finite element simulation of thermal barrier coating performance under thermal cycling. Surf Coat Technol 205:2042–2050
10. Hermosilla U, Karunaratne MSA, Jones IA, Hyde TH, Thomson RC (2009) Modelling the high temperature behaviour of TBCs using sequentially coupled microstructural–mechanical FE analyses. Mater Sci Eng A 513–514:302–310
11. Białas M (2008) Finite element analysis of stress distribution in thermal barrier coatings. Surf Coat Technol 202:6002–6010
12. Bednarz P (2007) Finite element simulation of stress evolution in thermal barrier coating systems. Ph.D. thesis, Forschungszentrum Jülich GmbH, Germany
13. Martena M, Botto D, Fino P, Sabbadini S, Gola MM, Badini C (2006) Modelling of TBC system failure: stress distribution as a function of TGO thickness and thermal expansion mismatch. Eng Fail Anal 13:409–426

14. Brodin H, Jinnestrand M, Sjöström S (2004) Modelling and experimental verification of delamination crack growth in an air-plasma-sprayed thermal barrier coating. In: 15th European Conference of Fracture (ECF15), Stockholm, Sweden
15. Nair BG, Singh JP, Grimsditch M (2004) Stress analysis in thermal barrier coatings subjected to long-term exposure in simulated turbine conditions. J Mater Sci 39:2043–2051
16. He MY, Hutchinson JW, Evans AG (2003) Simulation of stresses and delamination in a plasma-sprayed thermal barrier system upon thermal cycling. Mater Sci Eng A 345:172–178
17. Evans HE (2011) Oxidation failure of TBC systems: an assessment of mechanisms. Surf Coat Technol 206:1512–1521
18. Vaßen R, Giesen S, Stöver D (2009) Lifetime of plasma-sprayed thermal barrier coatings: comparison of numerical and experimental results. J Therm Spray Technol 18(5–6):835–845
19. Ghafouri-Azar R, Mostaghimi J, Chandra S (2006) Modeling development of residual stresses in thermal spray coatings. Comput Mater Sci 35:13–26
20. Klusemann B, Denzer R, Svendsen B (2012) Microstructure-based modeling of residual stresses in WC-12Co-sprayed coatings. J Therm Spray Technol 21(1):96–107
21. Nordhorn C, Mücke R, Vaßen R (2014) Simulation of the effect of realistic surface textures on thermally induced topcoat stress fields by two-dimensional interface functions. Surf Coat Technol 258:181–188, http://dx.doi.org/10.1016/j.surfcoat.2014.09.032
22. Gupta M, Skogsberg K, Nylén P (2014) Influence of topcoat-bondcoat interface roughness on stresses and lifetime in thermal barrier coatings. J Therm Spray Technol 23(1–2):170–181
23. Curry N, Markocsan N, Östergren L, Li X-H, Dorfman M (2013) Evaluation of the lifetime and thermal conductivity of dysprosia-stabilized thermal barrier coating systems. J Therm Spray Technol 22(6):864–872

Chapter 7
Modelling of Oxide Growth in TBCs

Visualisation of TGO growth in situ by experimental methods is highly non-trivial. One of the destructive ways of capturing the TGO growth by experiments is to treat several samples of the same coating for different operating times until failure and then inspecting the cross-sectional microstructure. However, in this case the TGO growth cannot be analysed at the same location in the TBC sample as the area inspected will differ between the samples and may not contain exactly the same surface topography. A non-destructive way could be to perform micro-computed tomography (Micro-CT) on a sample in several stages over the sample's lifetime. This technique, apart from being extremely expensive and time consuming, can be performed usually only on a very small sample with relatively low resolution. Thermography has been successfully performed in the past to analyse the crack growth during lifetime testing, but a low resolution is obtained with this technique [1]. High temperature X-ray diffraction (XRD) was used by Czech et al. to determine the thickness of TGO in situ under isothermal conditions by measuring the intensity curves, but the TGO profile could not be detected with this technique [2].

As apparent from the above discussion, modelling of TGO formation would be highly beneficial since using experimental methods would be highly challenging. The conventional way of modelling TGO growth (d_{TGO}) as a function of time (t) and temperature (T) is described a parabolic growth law given as [3]

$$d_{TGO} = A_{TGO} \bullet t^p \bullet \exp\left(-E_{TGO} / k_B T\right) \tag{7.1}$$

where A_{TGO} is TGO growth coefficient, p is TGO growth exponent, E_{TGO} is TGO growth activation energy and k_B is the Boltzmann constant. However, this parabolic growth law does not include the effect of bondcoat roughness on TGO formation. TGO growth models based on diffusion could thus provide a better reflection of the reality.

© The Author(s) 2015
M. Gupta, *Design of Thermal Barrier Coatings*, SpringerBriefs in Materials,
DOI 10.1007/978-3-319-17254-5_7

7.1   Diffusion-Based Modelling

7.1.1   Previous Work

A few studies have been performed earlier to model the TGO layer formation based
on diffusion of species. A multi-scale continuum mechanics approach based on a
coupled diffusion-constitutive framework was implemented by Busso et al. to study the
local stresses induced in TBCs using a parametric unit cell finite element model [4].
A numerical model describing the TGO growth by an oxygen diffusion–reaction
model was developed by Hille et al. to perform an analysis of a representative TBC
system subjected to a thermal cycling process and to make a parametric study for
different fracture strengths of the topcoat to determine its influence on the durability
of a TBC system [5]. A one-dimensional oxidation–diffusion model considering
both surface oxidation and coating–substrate interdiffusion as well as aluminium
depletion during operating conditions was developed by Yuan et al. to predict the
lifetime of TBCs [6]. A diffusion–reaction of aluminium and oxygen to form TGO
in TBCs was studied through an analytical model by Osorio et al. and the results
were compared with experiments to evaluate the TGO growth rate [7]. However, all
of the models discussed above did not incorporate the inherent TBC microstructure
or the topcoat–bondcoat interface topography.

Apart from achieving fundamental understanding of the TGO growth phenomena,
the development of TGO in situ during thermal cyclic loading conditions needs to
be implemented in a diffusion-based model to evaluate stresses developed due to
TGO formation. The following general assumptions based on the references above
could be made for developing a simplified TGO growth model:

1. TGO growth occurs only during the dwell phase of the lifetime testing.
2. The bondcoat mainly consists of aluminium and the TGO formed consists of
   pure alumina for a NiCoCrAlY bondcoat. Formation of other oxides can be
   neglected.
3. The reaction rate of aluminium and oxygen to form alumina is very high compared
   to the diffusion rates of the constituents.
4. The diffusion rate of oxygen within the topcoat can be considered to be very high
   since zirconia is transparent to oxygen flow as discussed in Sect. 3.5.

7.1.2   Real Interface Roughness Modelling

A 2D diffusion-based TGO growth model developed by the author using a computational fluid dynamics (CFD) approach consisting of real topcoat–bondcoat interface
topography extracted from cross-sectional micrographs based on the modelling
assumptions stated above is described here [8].

The cross-sectional microstructure images of TBC samples at failure were used to
define the initial geometry for the model. The microstructure images were imported

in a CAD software and a spline profile was carefully created over the image to capture the TGO profile. The upper edge of the TGO layer was defined as the initial as-sprayed stage in the diffusion model since only the inward growth of TGO occurring due to the diffusion of oxygen within the TGO towards the TGO–bondcoat interface was considered in the model for simplicity, and the lower edge was defined as the TGO thickness profile at the failure stage. As several cracks form within the TGO during testing and thus appear in the images due to the intense cracking leading to failure, the profiles were created where the TGO profile would have been before failure occurred based on prior experience. One of the imported microstructure images taken from a sample is shown in Fig. 7.1 along with the spline profiles. Bondcoat and topcoat layers were added to the spline profile to create a 2D planar model by extruding the structure shown with solid lines in Fig. 7.1. The porosity network and oxide inclusions within the topcoat and bondcoat layers were omitted from the simulation. Several such models need to be analysed for each coating for statistical significance. In this way the model could capture the real interface profile extracted from the microstructure images.

The TGO growth model was made in ANSYS Fluent 14.5 (ANSYS, Inc., Canonsburg, PA, USA) using a CFD approach. The generic convection–diffusion equation for additional scalars was used as the modelling concept, where the convective transport was omitted from the system.

A finite element mesh was created over the model. Fine elements were created near the topcoat–bondcoat interface to capture the TGO growth accurately while coarser elements were created near the outer edges. Two solid regions were defined in the system, one being the quiescent gas region containing oxygen and representing the topcoat and the other being the solid aluminium region representing the bondcoat.

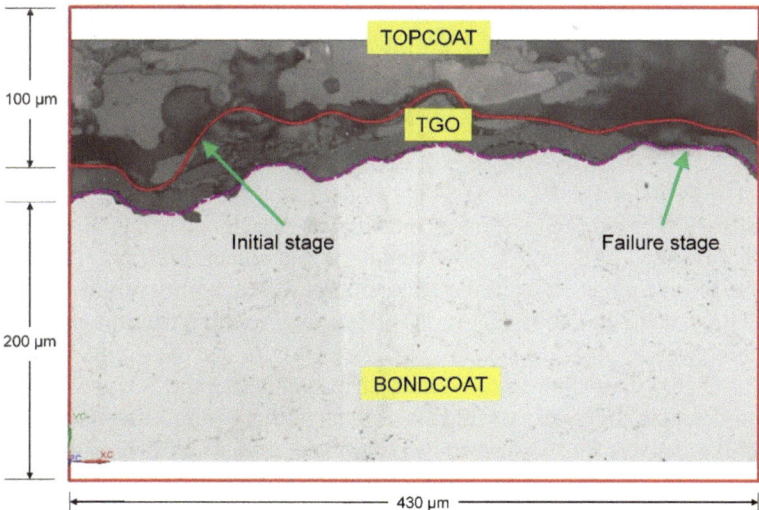

Fig. 7.1 TGO profile drawn over the microstructure image taken from one of the samples; the *solid lines* indicate the 2D model used in the TGO growth model [8]

The system consisted of the two initial solid scalars, oxygen and aluminium, transported by diffusion in the solid regions and a third solid scalar, alumina, formed during the simulation. As the oxygen and aluminium scalars congregate by diffusional transport, the alumina layer was produced.

An image showing the formed alumina concentration was saved after each time step. The TGO profiles at 50 % concentration of alumina were saved as a list of coordinates after several stages. These stages were selected so as to capture a significant growth of the TGO from one stage to the next stage. However, the number of time steps between these stages was kept the same for all the analysed models. The concentration of alumina at which these profiles were saved did not affect the final results. It must be noted here that since the TGO growth rate depends mainly on the diffusion coefficients, the approximate TGO thickness in different models at different stages would be in the same range; however, the TGO profile would be different which would depend on the initial profile of the topcoat–bondcoat interface. The TGO profiles extracted from the model in the form of coordinate files were later exported to NX for generating the TGO profile so as to capture the TGO growth. The timescale in the simulation was not considered to be relevant as the objective was to extract the TGO growth profile at different stages until failure.

Figure 7.2 shows three images selected from the ones saved after each time step in the TGO growth model for one of the profiles from the sample shown in Fig. 7.1; the scale bar illustrates the alumina concentration in fraction, for example, 50 % = 0.5. Figure 7.2a shows the initial profile at the beginning of the simulation where no alumina can be observed. Figure 7.2b shows the intermediate stage where the maximum alumina concentration can be observed to be around 50 % and very thin layer of TGO could be observed while Fig. 7.2c shows the mature stage of the TGO growth where a fairly thick TGO could be observed, considering the 50 % concentration level as indicated in the figure. The inward growth of TGO towards the bondcoat near the topcoat–bondcoat interface can be clearly observed in the images.

7.1.3 Recent Work

A 2D diffusion-based TGO growth model was developed using the modelling approach described in Sect. 7.1.2 by the author in a recent work [8]. The objective of this work was to make a tentative evaluation of the capability of the combination of the TGO growth and stress analysis models to assess the lifetime of TBCs in a comparative way.

The TGO growth model was first validated by comparing the TGO profiles created by the model with thermally cycled specimens with varying interface roughness. A qualitative analysis was performed by plotting the profiles created by the model for 50 % alumina concentration at different stages over the microstructure image at failure and comparing them with the initial and final stages at failure.

Figure 7.3 shows the profiles generated by the model plotted over the microstructure image at failure for the coating sample shown in Fig. 7.1. The initial as-sprayed and

Fig. 7.2 Images saved from the TGO growth model showing the (**a**) initial, (**b**) intermediate and (**c**) mature stages of TGO growth [8]

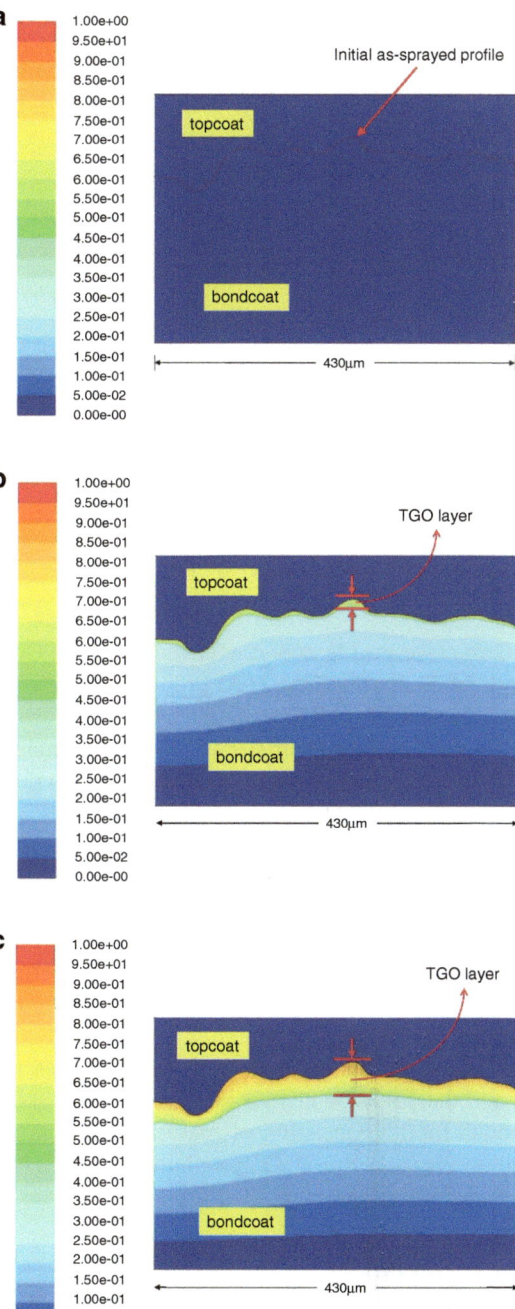

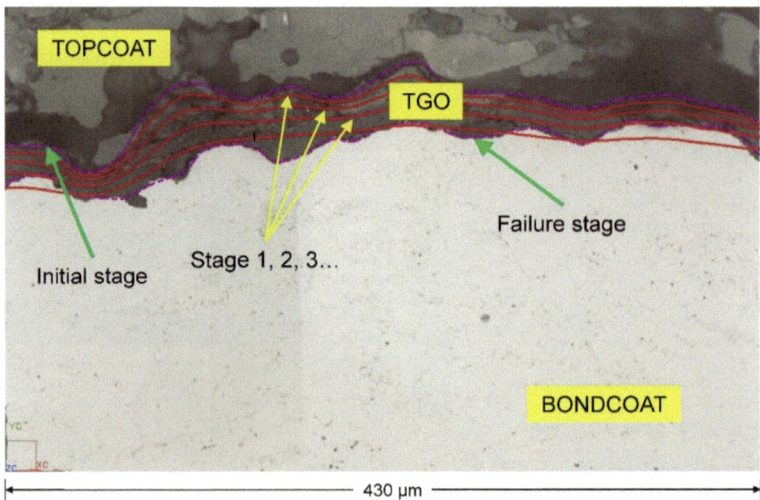

Fig. 7.3 TGO profiles generated from the TGO growth model indicated by *solid lines* plotted over the microstructure image at failure shown in Fig. 7.1 [8]

failure stage profile drawn over the image in CAD software is indicated by dashed lines while the different TGO growth stages extracted from the TGO growth model are indicated by solid lines. It can be observed that the generated TGO profiles closely follow the TGO profile in the microstructure image, and the thickness of the generated TGO seems to be similar to the actual TGO in the microstructure image.

Thereafter, the TGO profiles at different stages of TGO growth were extracted from the TGO growth model to the finite element model described in Sect. 6.2 for stress analysis and the time to stress inversion was evaluated. Three experimental specimens consisting of the same chemistry but with different topcoat–bondcoat interface roughness were studied by the models and the time to stress inversion was compared with the lifetimes measured experimentally. It was observed that the stress analysis model provided the same ranking for lifetime according to the time to stress inversion to the three samples as provided by the experiments.

Thus, the combination of the two models presented an effective approach to assess the stress behaviour and lifetime of TBCs in a comparative way. The TGO growth model consisting of real interface topography can be used as an effective tool to visualise the TGO growth in situ which could be otherwise a non-trivial task if performed experimentally.

7.2 Mixed Oxide Growth

As discussed in Sect. 3.5, the growth of mixed oxides in TBCs during operating conditions plays a major role in the formation of thermo-mechanical stresses. An attempt was thus made by the author in a recent work to understand the driving force

for cracking of CSN clusters and their role in TBC spallation in a better way [9].
The purpose of the study was to identify during which step of formation the CSN
cluster cracks and the driving mechanisms for cracking.

First, nanoindentation was performed on a CSN cluster in a TBC sample to map
Young's modulus of the various oxides included in CSN clusters. The values from
nanoindentation experiments were used to produce a Young's modulus map by assign-
ing appropriate Young's modulus to each zone as shown in Fig. 7.4 after making some
simplifications, such as removing porosity and very small material zones.

A finite element analysis was then performed using the established Young's
moduli and the exact geometry of the CSN cluster derived from micrographs using

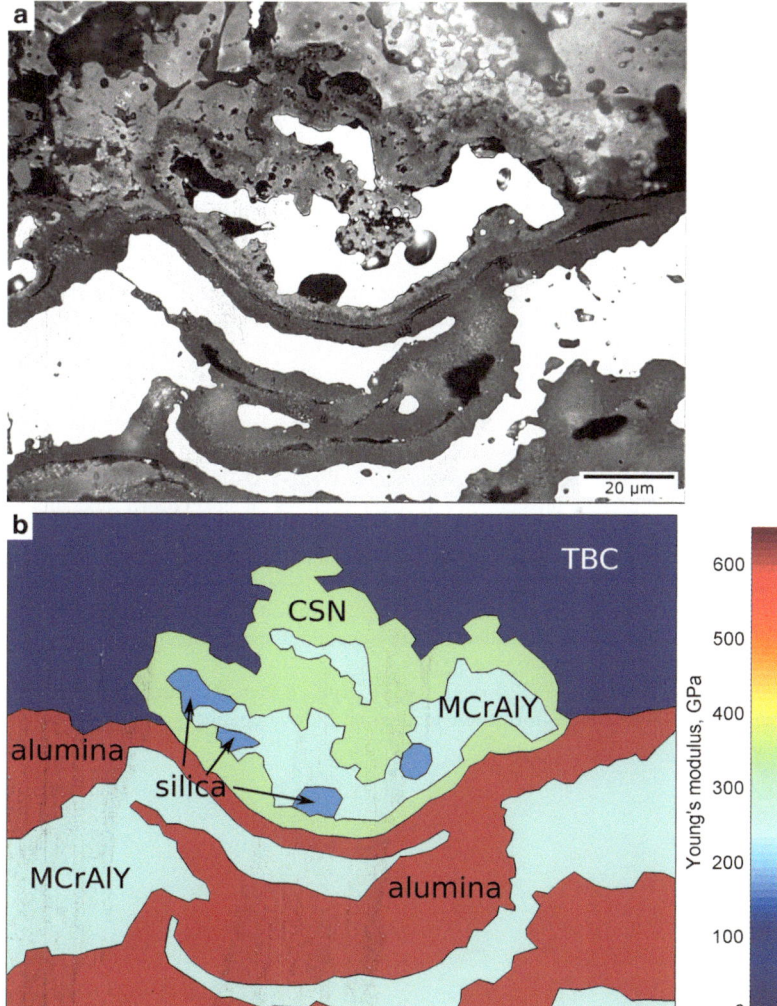

Fig. 7.4 (**a**) Light-optic image of an oxide cluster and (**b**) the corresponding Young's modulus
map, created from the values based on nanoindentation experiments after the material had been
identified in each part of the image through microscopy [9]

the model described in Sect. 6.2.1. Several stages of the CSN cluster formation process, as well as the volumetric increase associated with the oxidation of the last Ni-rich core, were modelled.

It was concluded from the modelling results that crack formation in the oxide clusters can occur due to the induced tensile stresses during cooling in the NiO core. On cooling, the high CTE of NiO induces tensile stresses in the NiO core and crack formation occurs in the cluster centre. The stresses introduced due to the oxidation of the Ni core were observed to relax fast enough at high temperature. Silica particles in the cluster could possibly assist cracking due to the CTE mismatch but is not the key factor in oxide cluster cracking.

References

1. Schweda M, Beck T, Singheiser L (2012) Thermal cycling damage evolution of a thermal barrier coating and the influence of substrate creep, interface roughness and pre-oxidation. Int J Mater Res 103:40–49
2. Czech N, Juez-Lorenzo M, Kolarik V, Stamma W (1998) Influence of the surface roughness on the oxide scale formation on MCrAlY coatings studied in situ by high temperature X-ray diffraction. Surf Coat Technol 108–109:36–42
3. Vaßen R, Giesen S, Stöver D (2009) Lifetime of plasma-sprayed thermal barrier coatings: comparison of numerical and experimental results. J Therm Spray Technol 18(5–6):835–845
4. Busso EP, Lin J, Sakurai S, Nakayama M (2001) A mechanistic study of oxidation-induced degradation in a plasma-sprayed thermal barrier coating system. Part I: model formulation. Acta Mater 49:1515–1528
5. Hille TS, Turteltaub S, Suiker ASJ (2011) Oxide growth and damage evolution in thermal barrier coatings. Eng Fract Mech 78:2139–2152
6. Yuan K, Eriksson R, Peng RL, Li X-H, Johansson S, Wang Y-D (2013) Modeling of microstructural evolution and lifetime prediction of MCrAlY coatings on nickel based superalloys during high temperature oxidation. Surf Coat Technol 232:204–215
7. Osorio JD, Giraldo J, Hernández JC, Toro A, Hernández-Ortiz JP (2014) Diffusion–reaction of aluminum and oxygen in thermally grown Al_2O_3 oxide layers. Heat Mass Transf 50:483–492
8. Gupta M, Eriksson R, Sand U, Nylén P (2014) A diffusion-based oxide layer growth model using real interface roughness in thermal barrier coatings for lifetime assessment. Surf Coat Technol. doi: 10.1016/j.surfcoat.2014.12.043
9. Eriksson R, Gupta M, Broitman E, Jonnalagadda P, Nylén P, Peng RL. Stress and cracking during Chromia-Spinel-NiO cluster formation in thermal barrier coating systems. J Therm Spray Technol. Submitted

Chapter 8
Conclusions: How to Design TBCs?

A two-step approach can be undertaken to design an optimised TBC by modelling:

1. *Design microstructure defects in topcoat*

The microstructure defects in the topcoat can be designed to reduce thermal conductivity and enhance strain tolerance of the topcoat by optimising the shape and size of these defects.

An automated procedure was described in this study using finite element modelling to predict thermal conductivity and Young's modulus of TBC topcoats. By comparing the predicted values with the experimental results, it was shown that finite element modelling approach utilising real microstructure images can be used as a powerful tool to predict and derive microstructure–property relationships.

It was shown that coating morphologies with similar microstructural features as the real microstructure images taken using SEM can be artificially generated by using an artificial coating morphology generator.

The presence of large globular pores with connected cracks in the coating microstructure was shown to significantly enhance the performance of TBCs. Low thermal conductivity, low Young's modulus and high lifetime were exhibited by these coatings. This relationship was derived by designing the microstructure using the combination of finite element modelling and artificial coating morphology generator and then verified experimentally.

2. *Design topcoat–bondcoat interface*

The topcoat–bondcoat interface in TBCs can be designed to increase the time to stress inversion, thus enhancing lifetime due to slower crack propagation by optimising the interface topography and oxide formation.

The stress inversion theory established using simplified sinusoidal curves was verified in this study by both 2D and 3D simulations using real topographies. It was shown that the modelling approach using real topographies can be a useful tool to attain insight into the stress distribution in topcoat–bondcoat interfaces. It was shown that the time to stress inversion can be used as an indicator to assess coating lifetime.

© The Author(s) 2015
M. Gupta, *Design of Thermal Barrier Coatings*, SpringerBriefs in Materials,
DOI 10.1007/978-3-319-17254-5_8

The results indicated that an optimal bondcoat topography could be designed through this technique.

The diffusion-based TGO growth model consisting of real interface topography developed in this work can be used as an effective tool to visualise the TGO growth in situ which could be otherwise a non-trivial task if performed experimentally. The combination of the TGO growth and stress analysis models developed in this work could be used as an effective approach to assess the stress behaviour and lifetime of TBCs in a comparative way. Based on the fundamental knowledge gained from these models about the relationships between topcoat–bondcoat interface roughness, TGO growth and lifetime of TBCs, the models need to be developed further to realise an optimised interface design.

The finite element model based on analysing real microstructure images was successfully implemented to achieve fundamental understanding of the phenomena of stress development and cracking during mixed oxide cluster formation in TBCs. It was concluded that the major reason for cracking was the large CTE between NiO and other oxides formed during thermal cycling.

Thus, it can be concluded from the results that the modelling approach described in this study could be used as a powerful tool to design new coatings as well as to achieve optimised microstructures and topographies which could significantly enhance the performance of TBCs.

8.1   Future Work

Further work in this field could be to develop the TGO growth model from 2D to 3D as it would provide a global view of TGO growth as well as enable to perform stress analysis in 3D. The inclusion of aluminium depletion and mixed oxides in the TGO growth model in the future would be highly valuable.

In recent years, coatings produced by SPS/SPPS are becoming of major interest. The analysis of failure mechanisms as well as the application of the models developed in this work to SPS/SPPS coatings could be another possibility in the future.